Y0-BDU-842

THE
CARUS MATHEMATICAL MONOGRAPHS

Published by
THE MATHEMATICAL ASSOCIATION OF AMERICA

Publication Committee

THE CARUS MATHEMATICAL MONOGRAPHS are an expression of the desire of Mrs. Mary Hegeler Carus, and of her son, Dr. Edward H. Carus, to contribute to the dissemination of mathematical knowledge by making accessible at nominal cost a series of expository presentations of the best thoughts and keenest researches in pure and applied mathematics. The publication of the first four of these monographs was made possible by a notable gift to the Mathematical Association of America by Mrs. Carus as sole trustee of the Edward C. Hegeler Trust Fund. The sales from these have resulted in the Carus Monograph Fund, and the Mathematical Association has used this as a revolving book fund to publish the fifth and sixth monographs.

The expositions of mathematical subjects which the monographs contain are set forth in a manner comprehensible not only to teachers and students specializing in mathematics, but also to scientific workers in other fields, and especially to the wide circle of thoughtful people who, having a moderate acquaintance with elementary mathematics, wish to extend their knowledge without prolonged and critical study of the mathematical journals and treatises. The scope of this series includes also historical and biographical monographs.

The following books in this series have been published to date:

No. 1. *Calculus of Variations,* by GILBERT AMES BLISS.

No. 2. *Analytic Functions of a Complex Variable,* by DAVID RAYMOND CURTISS

No. 3. *Mathematical Statistics,* by HENRY LEWIS RIETZ.

No. 4. *Projective Geometry,* by JOHN WESLEY YOUNG.

No. 5. *A History of Mathematics in America Before 1900,* by DAVID EUGENE SMITH and JEKUTHIEL GINSBURG.

No. 6. *Fourier Series and Orthogonal Polynomials* by DUNHAM JACKSON.

No. 7. *Vectors and Matrices,* by C. C. MACDUFFEE.

The Carus Mathematical Monographs

NUMBER THREE

MATHEMATICAL STATISTICS

By

HENRY LEWIS RIETZ

Professor of Mathematics, The University of Iowa

Published for

THE MATHEMATICAL ASSOCIATION OF AMERICA

by

THE OPEN COURT PUBLISHING COMPANY

LA SALLE • ILLINOIS

Published April 1927
Second Printing 1929
Third Printing 1936
Fourth Printing 1943
Fifth Printing 1947

Reprinted by
JOHN S. SWIFT CO., INC.
CHICAGO
ST. LOUIS CINCINNATI NEW YORK

PREFACE

This book on mathematical statistics is the third of the series of Carus Mathematical Monographs. The purpose of the monographs, admirably expressed by Professor Bliss in the first book of the series, is "to make the essential features of various mathematical theories more accessible and attractive to as many persons as possible who have an interest in mathematics but who may not be specialists in the particular theory presented."

The problem of making statistical theory available has been changed considerably during the past two or three years by the appearance of a large number of text-books on statistical methods. In the course of preparation of the manuscript of the present volume, the writer felt at one time that perhaps the recent books had covered the ground in such a way as to accomplish the main purposes of the monograph which was in process of preparation. But further consideration gave support to the view that although the recent books on statistical method will serve useful purposes in the teaching and standardization of statistical practice, they have not, in general, gone far toward exposing the nature of the underlying theory, and some of them may even give misleading impressions as to the place and importance of probability theory in statistical analysis.

It thus appears that an exposition of certain essential features of the theory involved in statistical analysis would conform to the purposes of the Carus Mathematical Monographs, particularly if the exposition could be made interesting to the general mathematical reader. It is not the intention in the above remarks to imply a criticism of the books in question. These books serve certain useful purposes. In them the emphasis has been very properly placed on the use of devices which facilitate the description and analysis of data.

The present monograph will accomplish its main purpose if it makes a slight contribution toward shifting the emphasis and point of view in the study of statistics in the direction of the consideration of the underlying theory involved in certain highly important methods of statistical analysis, and if it introduces some of the recent advances in mathematical statistics to a wider range of readers. With this as our main purpose it is natural that no great effort is being made to present a well-balanced discussion of all the many available topics. This will be fairly obvious from omissions which will be noted in the following pages. For example, the very important elementary methods of description and analysis of data by purely graphic methods and by the use of various kinds of averages and measures of dispersion are for the most part omitted owing to the fact that these methods are so available in recent elementary books that it seems unnecessary to deal with them in this monograph. On the other hand, topics which suggest making the underlying theories more available are emphasized.

For the purpose of reaching a relatively large number of readers, we are fortunate in that considerable portions of the present monograph can be read by those who have relatively little knowledge of college mathematics. However, the exposition is designed, in general, for readers of a certain degree of mathematical maturity, and presupposes an acquaintance with elementary differential and integral calculus, and with the elementary principles of probability as presented in various books on college algebra for freshmen.

A brief list of references is given at the end of Chapter VII. This is not a bibliography but simply includes books and papers to which attention has been directed in the course of the text by the use of superscripts.

The author desires to express his special indebtedness to Professor Burton H. Camp who read critically the entire manuscript and made many valuable suggestions that resulted in improvements. The author is also indebted to Professor A. R. Crathorne for suggestions on Chapter I and to Professor E. W. Chittenden for certain suggestions on Chapters II and III. Lastly, the author is deeply indebted to Professor Bliss and to Professor Curtiss of the Publication Committee for important criticisms and suggestions, many of which were made with special reference to the purposes of the Carus Mathematical Monographs.

HENRY L. RIETZ

THE UNIVERSITY OF IOWA
December, 1926

TABLE OF CONTENTS

CHAPTER I

THE NATURE OF THE PROBLEMS AND UNDERLYING CONCEPTS OF MATHEMATICAL STATISTICS

1. The scope of mathematical statistics. The bounds of mathematical statistics are not sharply defined. It is not uncommon to include under mathematical statistics such topics as interpolation theory, approximate integration, periodogram analysis, index numbers, actuarial theory, and various other topics from the calculus of observations. In fact, it seems that mathematical statistics in its most extended meaning may be regarded as including all the mathematics applied to the analysis of quantitative data obtained from observation. On the other hand, a number of mathematicians and statisticians have implied by their writings a limitation of mathematical statistics to the consideration of such questions of frequency, probability, averages, mathematical expectation, and dispersion as are likely to arise in the characterization and analysis of masses of quantitative data. Borel has expressed this somewhat restricted point of view in his statement[1] that the general problem of mathematical statistics is to determine a system of drawings carried out with urns of fixed composition, in such a way that the results of a series of drawings lead, with a very high degree of probability, to a table of values identical with the table of observed values.

[1] For footnote references, see pp. 173–77.

On account of the different views concerning the boundaries of the field of mathematical statistics there arose early in the preparation of this monograph questions of some difficulty in the selection of topics to be included. Although no attempt will be made here to answer the question as to the appropriate boundaries of the field for all purposes, nevertheless it will be convenient, partly because of limitations of space, to adopt a somewhat restricted view with respect to the topics to be included. To be more specific, the exposition of mathematical statistics here given will be limited to certain methods and theories which, in their inception, center around the names of Bernoulli, De Moivre, Laplace, Lexis, Tchebycheff, Gram, Pearson, Edgeworth, and Charlier, and which have been much developed by other contributors. These methods and theories are much concerned with such concepts as frequency, probability, averages, mathematical expectation, dispersion, and correlation.

2. **Historical remarks.** While we are currently experiencing a period of special activity in mathematical statistics which dates back only about forty years, some of the concepts of mathematical statistics are by no means of recent origin. The word "statistics" is itself a comparatively new word as shown by the fact that its first occurrence in English thus far noted seems to have been in J. F. von Bielfeld, *The Elements of Universal Erudition*, translated by W. Hooper, London, 1770. Notwithstanding the comparatively recent introduction of the word, certain fundamental concepts of mathematical statistics to which attention is directed in this monograph date back to the first publication relating to Bernoulli's theorem in 1713. The line of development started by Bernoulli was carried

forward by Stirling (1730), De Moivre (1733), Euler (1738), and Maclaurin (1742), and culminated in the formulation of the probability theory of Laplace. The *Théorie Analytique des Probabilités* of Laplace published in 1812 is the most significant publication underlying mathematical statistics. For a period of approximately fifty years following the publication of this monumental work there was relatively little of importance contributed to the subject. While we should not overlook Poisson's extension of the Bernoulli theory to cases where the probability is not constant, Gauss's development of methods for the adjustment of observations, Bravais's extension of the normal law to functions of two and three variables, Quetelet's activities as a popularizer of social statistics, nevertheless there was on the whole in this period of fifty years little progress.

The lack of progress in this period may be attributed to at least three factors: (1) Laplace left many of his results in the form of approximations that would not readily form the basis for further development; (2) the followers of Gauss retarded progress in the generalization of frequency theory by overpromoting the idea that deviations from the normal law of frequency are due to lack of data; (3) Quetelet overpopularized the idea of the stability of certain striking forms of social statistics, for example, the stability of the number of suicides per year, with the natural result that his activities cast upon statistics a suspicion of quackery which exists even to some extent at present.

An important step in advance was taken in 1877 in the publication of the contributions of Lexis to the classification of statistical distributions with respect to normal, supernormal, and subnormal dispersion. This theory will receive attention in the present monograph.

The development of generalized frequency curves and the contributions to a theory of correlation from 1885 to 1900 started the period of activity in mathematical statistics in which we find ourselves at present. The present monograph deals largely with the progress in this period, and with the earlier underlying theory which facilitated relatively recent progress.

3. **Two general types of problems.** For purposes of description it seems convenient to recognize two general classes of problems with which we are concerned in mathematical statistics. In the problems of the first class our concern is largely with the characterization of a set of numerical measurements or estimates of some attribute or attributes of a given set of individuals. For example, we may establish the facts about the heights of 1,000 men by finding averages, measures of dispersion, and various statistical indexes. Our problem may be limited to a characterization of the heights of these 1,000 men.

In the problems of the second class we regard the data obtained from observation and measurement as a random sample drawn from a well-defined class of items which may include either a limited or an unlimited supply. Such a well-defined class of items may be called the "population" or universe of discourse. We are in this case concerned with using the properties of a random sample of variates for the purpose of drawing inferences about the larger population from which the sample was drawn. For example, in this class of problems involving the heights of the 1,000 men we would be concerned with the question: What approximate or probable inferences may be drawn about the statures of a whole race of men from an analysis of the heights of a sample of 1,000 men drawn at random from the men of the race? In dealing with such questions, we should in the first place consider the difficulties involved in drawing a sample that is truly random, and in the next place the problem of developing certain parts of the theory of probability involved in statistical inference.

The two classes of problems to which we have directed attention are not, however, entirely distinct with regard to their treatment. For example, the conceptions of probable and standard error may be used both in describing the facts about a sample and in indicating the probable degree of precision of inferences which go beyond the observed sample by dealing with certain properties of the population from which we conceive the sample to be drawn. Moreover, a satisfactory description of a sample is not likely to be so purely descriptive as wholly to prevent the mind from dwelling on the inner meaning of the facts in relation to the population from which the sample is drawn.

As a preliminary to dealing in later chapters with certain of the problems falling under these two general classes we shall attempt in the present chapter to discuss briefly the nature of certain underlying concepts. We shall find it convenient to consider these concepts in pairs as follows: relative frequency and probability; observed and theoretical frequency distributions; arithmetic mean and mathematical expectation; mode and most probable value; moments and mathematical expectations of a power of a variable.

4. Relative frequency and probability. The frequency f of the occurrence of a character or event among s possible occurrences is one of the simplest items of statistical information. For example, any one of the following items illustrates such statistical information: Five deaths in a year among 1,000 persons aged 30, nearest birthday; 610 boys among the last 1,200 children born in a city; 400 married men out of a total of 1,000 men of age 23; twelve cases of 7 heads in throwing 7 coins 1,536 times.

The determination of the numerical values of the relative frequencies f/s corresponding to such items is one of the simplest problems of statistics. This simple problem suggests a fundamental problem concerning the probable or expected values of such relative frequencies if s were a very large number. When s is a large number, the relative frequency f/s is very commonly accepted in applied statistics as an approximate measure of the probability of occurrence of the event or character on a given occasion.

To take an illustration from an important statistical problem, let us assume that among l persons equally likely to live a year we find d observed deaths during the year. That is, we assume that d represents the frequency of deaths per year among the l persons each exposed for one year to the hazards of death. If l is fairly large, the relative frequency d/l is often regarded as an approximation to what is to be defined as the probability of death of one such person within a year. In fact, it is a fundamental assumption of actuarial science that we may regard such a relative frequency as an approximation to the probability of death when a sufficiently large number of persons are exposed to the hazards of death. For a numerical illus tration, suppose there are 600 deaths among 100,000 persons exposed for a year at age 30. We accept .006 as an approximation to the probability in question at age 30. In the method of finding such an approximation we decide on a population which constitutes an appropriate class for investigation and in which individuals satisfy certain conditions as to likeness. Then we depend on observation to obtain the items which lead to the relative frequency which we may regard as an approximation to the probability.

For an ideal population, let us conceive an urn containing white and black balls alike except as to color and thoroughly mixed. Suppose further for the present that we do not know the ratio of the number of white balls to the total number in this urn which we may conceive to contain either any finite number or an indefinitely large number of balls. This ratio is often called the probability of drawing a white ball. When the number in the urn is finite, we make drawings at random consisting of s balls taken one at a time with replacements to keep the ratio of the numbers of white and black balls constant. If we may assume the number in the urn to be infinite, the drawings may under certain conditions be made without replacements. Suppose we obtain f white balls as a result of thus drawing s balls, then we say that f/s is the relative frequency with which we drew white balls. When s is large, this relative frequency would ordinarily give us an approximate value of the probability of drawing a white ball in one trial, that is, an approximate value of the ratio of white balls to the total number of balls in the urn.

Thus far we have not defined probability, but have

presented illustrations of approximations to probabilities. While these illustrations seem to suggest a definition, it is nevertheless difficult to frame a definition that is satisfactory and includes all forms of probability. The need for the concepts of relative frequency and probability in statistics arises when we are associating two events such that the first may be regarded as a *trial* and the second may be regarded as a *success* or a *failure* depending on the result of the trial. The relative frequency of success is then the ratio of the number of successes to the total number of trials.

If the relative frequency of success approaches a limit when the trial is repeated indefinitely under the same set of circumstances, this limit is called the probability of success in one trial.

There are some objections to this definition of probability as well as to any other that we could propose. One objection is concerned with questioning the validity of the assumption that a limit of the relative frequency exists, and another relates to the meaning of the expression, "the same set of circumstances." That the limit exists is an empirical assumption whose validity cannot be proved, but experience with data in many fields has given much support to the reasonableness and usefulness of the assumption. The objection based on the difficulty of controlling conditions so as to repeat the trial under the same set of circumstances is an objection that could be brought against experimental science in general with respect to the difficulties of repeating experiments under the same circumstances. The experiments are repeated as nearly as circumstance permits.

It seems fairly obvious that the development of statistical concepts is approached more naturally from this limit definition than from the familiar definitions suggested by games of chance. However, we shall at certain points in our treatment (for example, see § 11) give attention to the fact that various definitions of probability exist in which the assumptions differ from those involved in the above definition. The meaning of probability in statistics is fairly well expressed for some purposes by any one of the expressions, *theoretical relative frequency*, *presumptive relative frequency*, or *expected value* of a relative frequency. Indeed, we sometimes express the fact that the relative frequency f/s is assumed to have the probability p as a limit when $s \to \infty$ in abbreviated form by writing $E(f/s) = p$, where $E(f/s)$ is read, "expected value of f/s." It is fairly clear that in our definition of probability we simply idealize actual experience by assuming the existence of a limit of the relative frequency. This idealization, for purposes of definition, is in some respects analogous to the idealization of the chalk mark into the straight line of geometry.

In certain cases, notably in games of chance or urn schemata, the probability may be obtained without collecting statistical data on frequencies. Such cases arise when we have urn schemata of which we know the ratio of the number of white balls to the total number. For example, suppose an urn contains 7 white and 3 black balls and that we are to inquire into the probability that a ball to be drawn will be white. We could experiment by drawing one ball at a time with replacements until we had made a very large number of drawings and then estimate the probability from the ratio of the number of white balls to the total number of balls drawn. It would however in this case ordinarily be much more convenient and satisfying to examine the balls to note that they are alike except as to color and then make certain assumptions that would give us the probability without actually making the trials.

Thus, when all the possible ways of drawing the balls one at a time may be analyzed into 10 equally likely ways, and when 7 of these 10 ways give white balls, we assume that 7/10 is the probability that the ball to be drawn in one trial will be white. This simple case illustrates the following process of arriving at a probability:

If all of an aggregate of ways of obtaining successes and failures can be analyzed into s' possible mutually exclusive ways each of which is equally likely; and if f' of these ways give successes, the probability of a success in a single trial may be taken to be f'/s'.

Thus in throwing a single die, what is the probability of obtaining an ace? We assume that there are 6 equally likely ways in which the die may fall. One of these ways gives an ace. Hence, we say 1/6 is the probability of throwing an ace. A probability whose value is thus obtained from an analysis of ways of occurrence into sets of equally likely cases and a segregation of the cases in which a success would occur is sometimes called an *a priori probability*, while a probability whose approximate value is obtained from actual statistical data on repeated trials is called an *a posteriori* or *statistical probability*.

In making an analysis to study probabilities, difficult questions arise both as to the meaning and fulfilment of the condition that the ways are to be "equally likely." These questions have been the subject of lively debates by mathematicians and philosophers since the time of Laplace. It has been fairly obvious that the expression "equally likely ways" implies as a necessary condition that we have no information leading us to expect the event to occur in one of two ways rather than in the other, but serious doubt very naturally arises as to the sufficiency of this condition. In fact, it is fairly clear that lack of information is not sufficient. For example, lack of information as to whether a spinning coin is symmetrical and homogeneous does not assist one in passing on the validity of the assumption that it is equally likely to turn up head or tail. It is when we have all available relevant information on such matters as symmetry and homogeneity that we have a basis for the inference that the two ways are equally likely, or not equally likely. Similarly, lack of information about two large groups of men of age 30 would not assist us in making the inference that the mortality rates or probabilities of death are approximately equal for the two groups. On the other hand, relevant information in regard to the results of recent medical examinations, occupations, habits, and family histories would give support to certain inferences or assumptions concerning the equality or inequality of the mortality rates for the two groups.

5. Observed and theoretical frequency distributions. In many statistical investigations, it is convenient to partition the whole group of observations into subgroups or classes so as to show the number or frequency of observations in each class. Such an exhibit of observations is called an "observed frequency distribution." As illustrations we present the following, where the rows marked F are the observed frequency distributions:

Example 1. A = lengths of ears of corn in inches.

A....	3	4.5	6.0	7.5	9.0	10.5	12.0
F....	1	3	20	63	170	67	3

Example 2. A = prices of commodities for 1919 relative to price of 1913 as a base.

A..	62	87	112	137	162	187	212	237	262	287	312	337	362	387	412	437	462
F..	1	2	5	16	39	66	61	36	38	24	9	3	3	3	0	0	1

Example 3. A = heights of men in inches.

A....	61	62	63	64	65	66	67	68	69	70	71	72	73	74
F....	2	10	11	38	57	93	106	126	109	87	75	23	9	4

In Example 1 the whole group of ears of corn is arranged in classes with respect to length of ears. The class interval in this case is taken to be one and one-half inches. In Example 2 the class interval is a number, twenty-five; in Example 3, it is one inch.

If the variable x takes values $x_1, x_2, \ldots, x_n$ with the corresponding probabilities $p_1, p_2, \ldots, p_n$, we call the system of values $x_1, x_2, \ldots, x_n$ and the associated probabilities or numbers proportional to them, the theoretical frequency distribution of the variable x. Thus, we may write for the theoretical frequency distribution of the number of heads in throwing three coins:

Heads..................	0	1	2	3
Probabilities...........	1/8	3/8	3/8	1/8
Theoretical frequencies..	1	3	3	1

When for a given set of values of a variable x there exists a function $F(x)$ such that the ratio of the number of values of x on the interval ab to the number on the interval $a'b'$ is the ratio of the integrals

$$\int_a^b F(x)dx : \int_{a'}^{b'} F(x)dx ,$$

for all choices of the intervals ab and $a'b'$, then $F(x)$ is called the *frequency function*, or the *probability density*, or the *law of distribution* of the values of x. The curve $y=F(x)$ is called a *theoretical frequency curve*, or more briefly the *frequency curve.*

To devise methods for the description and characterization of the various types of frequency distributions which occur in practical problems of statistics is clearly

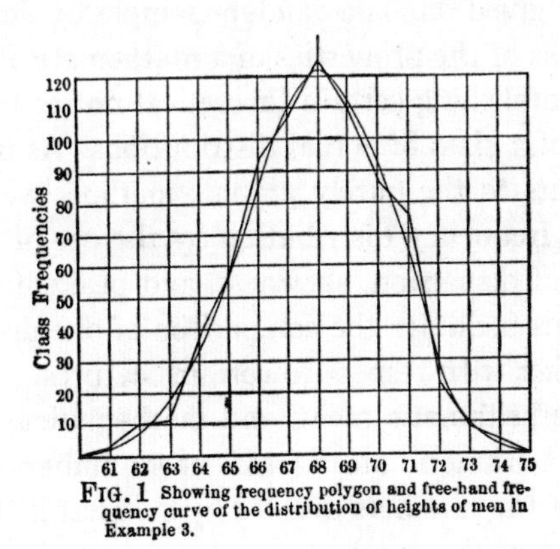

FIG. 1 Showing frequency polygon and free-hand frequency curve of the distribution of heights of men in Example 3.

of fundamental importance. Such a description or characterization may be effected with various degrees of refinement ranging all the way from one extreme with a simple frequency polygon or freehand curve (Fig. 1) representing frequencies by ordinates, to a description at the other extreme by means of a theoretical frequency curve grounded in the theory of probability.

It is fairly obvious that the latter type of description is likely to be much more satisfactory than the former because a deeper meaning is surely given to an observed distribution if we can effectively describe it by means of a theoretical frequency curve than if we can give only a freehand or an empirical curve as the approximate representation. However, we should not overlook the fact that the description by means of a theoretical curve may be too ponderous and laborious for the particular purpose of an analysis. Indeed, the use of the theoretical curve is likely to be justified in a large way only when it facilitates the study of the properties of the class of distributions of which the given one is a random sample by enabling us to make use of the properties of a mathematical function $F(x)$ in establishing certain theoretical norms for the description of a class of actual distributions. As important supplements to the purely graphic method, we may describe the frequency distribution by the use of averages, measures of dispersion, skewness, and peakedness. Such descriptions facilitate the comparison of one distribution with another with respect to certain features.

6. **The arithmetic mean and mathematical expectation.** The arithmetic mean (AM) of n numbers is simply the sum of the numbers divided by n. That is, the arithmetic mean of the numbers

$$x_1, x_2, \ldots, x_n$$

is given by the formula

$$AM = \frac{x_1 + x_2 + \cdots + x_n}{n} . \tag{1}$$

The AM is thus what is usually meant by the terms "mean," "average," or "mean value" when used without further qualification. If the values $x_1, x_2, \ldots, x_n$ occur with corresponding frequencies $f_1, f_2, \ldots, f_n$, respectively, where $f_1+f_2+\ldots+f_n=s$, then it follows from (1) that the arithmetic mean is given by

$$AM=\frac{f_1x_1+f_2x_2+\cdots+f_nx_n}{s}, \tag{2}$$

$$=\frac{f_1}{s}x_1+\frac{f_2}{s}x_2+\cdots+\frac{f_n}{s}x_n, \tag{3}$$

where

$$f_1/s+f_2/s+\cdots+f_n/s=1.$$

The arithmetic mean given by (2) is sometimes called a "weighted arithmetic mean" where $f_1, f_2, \ldots, f_n$ are the weights of the values $x_1, x_2, \ldots, x_n$, respectively, and (3) may similarly be regarded as a weighted arithmetic mean, where

$$f_1/s, f_2/s, \ldots, f_n/s$$

are the weights of $x_1, x_2, \ldots, x_n$, respectively.

For our present purpose it is important to note that the coefficients of $x_1, x_2, \ldots, x_n$ in (3) are the relative frequencies of occurrence of these values. By definition of statistical probabilities, the limiting value of f_t/s as s increases indefinitely is p_t, where p_t is the assumed probability of the occurrence of a value x_t among a set of mutually exclusive values $x_1, x_2, \ldots, x_n$. Hence, as the number of cases considered becomes infinite, the arithmetic mean would approach a value given by

$$AM=p_1x_1+p_2x_2+\cdots+p_nx_n, \tag{4}$$

where the probabilities $p_1, p_2, \ldots, p_n$ may be regarded as the weights of the corresponding values

$$x_1, x_2, \ldots, x_n.$$

The *mathematical expectation* of the experimenter or the *expected value* of the variable is a concept that has been much used by various continental European writers on mathematical statistics. Suppose we consider the probabilities $p_1, p_2, \ldots, p_n$ of n mutually exclusive events $E_1, E_2, \ldots, E_n$, so that $p_1+p_2+\cdots+p_n=1$. Suppose that the occurrence of one of these, say E_t, on a given occasion yields a value x_t of a variable x. Then the mathematical expectation or expected value $E(x)$ of the variable x which takes on values $x_1, x_2, \ldots x_n$ with the probabilities $p_1, p_2, \ldots, p_n$, respectively, may be defined as

$$E(x)=p_1x_1+p_2x_2+\cdots+p_nx_n. \tag{5}$$

We thus note by a comparison of (4) and (5) the identity of the limit of the mean value and the mathematical expectation.

Furthermore, in dealing with a theoretical distribution in which p_t is the probability that a variable x assumes a value x_t among the possible mutually exclusive values $x_1, x_2, \ldots, x_n$, and $p_1+p_2+\cdots+p_n=1$, we have

$$AM=p_1x_1+p_2x_2+\cdots+p_nx_n. \tag{6}$$

That is, the mathematical expectation of a variable x and its mean value from the appropriate theoretical distribution are identical. While there are probably differences of opinion as to the relative merits of the language involving mathematical expectation or expected value in comparison with the language which uses the mean value of a theoretical distribution, or mean value as the number of cases becomes infinite, the language of expectation seems the more elegant in many theoretical discussions. For the discussions in the present monograph we shall employ both of these types of language.

7. The mode and the most probable value. The *mode* or *modal* value of a variable is that value which occurs most frequently (that is, is most fashionable) if such a value exists.

Rough approximations to the mode are used considerably in general discourse. To illustrate, the meaning of the term "average" as frequently used in the newspapers in speaking of the average man seems to be a sort of crude approximation to the mode. That is, the term "average" in this connection usually implies a type which occurs oftener than any other single type.

The mode presents one of the most striking characteristics of a frequency distribution. For example, consider the frequency distribution of ears of corn with respect to rows of kernels on ears as given in following table:

A.....	10	12	14	16	18	20	22	24
F.....	1	16	109	241	235	116	41	10,

where A = number of rows of kernels and F = frequency. It may be noted that the frequency increases up to the class with 16 rows and then decreases. The mode in relation to a frequency distribution is a value to which there corresponds a greater frequency than to values just preceding or immediately following it in the arrangement. That is, the mode is the value of the variable for which the frequency is a maximum. A distribution may have more than one maximum, but the most common types of frequency distributions of both theoretical and practical interest in statistics will be found to have only one mode.

The expression "most probable value" of the number of successes in s trials is used in the general theory of probability for the number to which corresponds a larger probability of occurrence than to any other single number which can be named. For example, in throwing 100 coins, the most probable number of heads is 50, because 50 is more likely than any other single number.

This does not mean, however, that the probability of throwing exactly 50 heads is large. In fact, it is small, but nevertheless greater than the probability of throwing 49 or any other single number of heads. In other words, the most probable value is the modal value of the appropriate theoretical distribution.

8. **Moments and the mathematical expectations of powers of a variable.** With observed frequencies $f_1, f_2, \ldots, f_n$ corresponding to $x_1, x_2, \ldots, x_n$, respectively, and with $f_1+f_2+\cdots+f_n=s$, *the* kth *order moment, per unit frequency*, is defined as

$$\mu_k' = \frac{1}{s}\sum_{t=1}^{t=n} f_t x_t^k \,, \tag{7}$$

which is the arithmetic mean of the kth powers of the variates. For the sake of brevity, we shall ordinarily use the word "moment" as an abbreviation for "moment per unit frequency," when this usage will lead to no misunderstanding of the meaning.

Consider a theoretical distribution of a variable x taking values $x_t (t=1, 2, \ldots, n)$. Let the corresponding probabilities of occurrence $p_t (t=1, 2, \ldots, n)$ be represented as y-ordinates. Then the moment of order k of the ordinates about the y-axis is defined as

$$\mu_k' = \sum_{t=1}^{t=n} p_t x_t^k . \tag{8}$$

The mathematical expectation of the kth power of x is likewise defined as the second member of this equality so that the kth moment of the theoretical distribution and the mathematical expectation of the kth power of the variable x are identical.

When we have a theoretical distribution ranging from $x=a$ to $x=b$, and given by a frequency function (p. 13) $y=F(x)$, we write in place of (8)

$$\mu_k' = \int_a^b x^k F(x)dx ,$$

where $F(x)dx$ gives, to within infinitesimals of higher order, the probability that a value of x taken at random falls in any assigned interval x to $x+dx$.

When the axis of moments is parallel to the y-axis and passes through the arithmetic mean or centroid $\bar{x}$ of the variable x, the primes will be dropped from the μ's which denote the moments. Thus, we write

$$\mu_k = \frac{1}{s}\sum_{t=1}^{t=n} f_t(x_t - \bar{x})^k = \frac{1}{s}\sum_{t=1}^{t=n} f_t(x_t - \mu_1')^k , \tag{9}$$

where the arithmetic mean of the values of x is $\bar{x} = \mu_1'$.

The square root of the second moment μ_2 about the arithmetic mean is called the *standard deviation* and is very commonly denoted by σ. That is, the standard deviation is the root-mean-square of the deviations of a set of numbers from their arithmetic mean. In the language of mechanics, σ is the radius of gyration of a set of s equal particles, with respect to a given centroidal axis.

It is often important to be able to compute the moments about the axis through the centroid from those about an arbitrary parallel axis. For this purpose the following relations are easily established by expanding the binomial in (9) and then making some slight simplifications:

$$\mu_0 = \mu_0' = 1\,, \qquad \mu_1 = 0\,, \qquad \mu_2 = \mu_2' - \mu_1'^2\,,$$

$$\mu_3 = \mu_3' - 3\mu_1'\mu_2' + 2\mu_1'^3\,,$$

$$\mu_4 = \mu_4' - 4\mu_1'\mu_3' + 6\mu_1'^2\mu_2' - 3\mu_1'^4\,,$$

$$\cdot \quad \cdot \quad \cdot \quad \cdot \quad \cdot \quad \cdot \quad \cdot \quad \cdot \quad \cdot \quad \cdot \quad \cdot \quad \cdot$$

$$\mu_n = \sum_{i=0}^{n} \binom{n}{i} (-\mu_1')^i \mu_{n-i}'\,,$$

where

$$\binom{n}{i} = \frac{n!}{i!(n-i)!}$$

is the number of combinations of n things taken i at a time.

These relations are very useful in certain problems of practical statistics because the moments μ_k' ($k = 1, 2, \ldots$) are ordinarily computed first about an axis conveniently chosen, and then the moments μ_k about the parallel centroidal axis may be found by means of the above relations. In particular, $\mu_2 = \mu_2' - \mu_1'^2$ expresses the very important relation that the second moment μ_2 about the arithmetic mean is equal to the second moment μ_2' about an arbitrary origin diminished by the square $\mu_1'^2$ of the arithmetic mean measured from the arbitrary origin. This is a familiar proposition of elementary mechanics when the mean is replaced by the centroid.

When we pass from (9) to corresponding expectations, the relation $\mu_2 = \mu_2' - \mu_1'^2$, written in the form $\mu_2' = \mu_1'^2 + \mu_2$, tells us that the expected value, $E(x^2)$, of x^2 is equal to the square, $[E(x)]^2$, of the expected value of x increased by the expected value, $E\{[x - E(x)]^2\}$, of the square of the deviations of x from its expected value.

CHAPTER II

RELATIVE FREQUENCIES IN SIMPLE SAMPLING

9. The binomial description of frequency. In Chapter I attention was directed to the very simple process of finding the relative frequency of occurrence of an event or character among s cases in question. Let us now conceive of repeating the process of finding relative frequencies on many random samples each consisting of s items drawn from the same population. To characterize the degree of stability or the degree of dispersion of such a series of relative frequencies is a fundamental statistical problem.

To illustrate, suppose we repeat the throwing of a set of 1,000 coins many times. An observed frequency distribution could then be exhibited with respect to the number of heads obtained in each set of 1,000, or with respect to the relative frequency of heads in sets of 1,000. Such a procedure would be a laborious experimental treatment of the problem of the distribution of relative frequencies from repeated trials. What we seek is a mathematical method of obtaining the theoretical frequency distribution with respect to the number of heads or with respect to the relative frequency of heads in the sets.

To consider a more general problem, suppose we draw many sets of s balls from an urn one at a time with replacements, and let p be the probability of success in drawing a white ball in one trial. The problem we set is to determine the theoretical frequency distribution with respect to the number of white balls per set of s, or with respect to the relative frequency of white balls in the sets.

To consider this problem, let q be the probability of failure to draw a white ball in one trial so that $p+q=1$. Then the probabilities of exactly $m=0, 1, 2, \ldots, s$ successes in s trials are given by the successive terms of the binomial expansion

$$(1)\quad \begin{cases} (q+p)^s = q^s + spq^{s-1} + \binom{s}{2} p^2 q^{s-2} \\ \qquad + \cdots + \binom{s}{m} p^m q^{s-m} + \cdots + p^s \end{cases}$$

where

$$\binom{s}{m} = \binom{s}{s-m} = \frac{s!}{m!\,(s-m)!}\,.$$

Derivations of this formula for the probability of m successes in s trials from certain definitions of probability are given in books on college algebra for freshmen. For a derivation starting from the definition of probability as a limit, the reader is referred to Coolidge.[2] A frequency distribution with class frequencies proportional to the terms of (1) is sometimes called a Bernoulli distribution. Such a theoretical distribution shows not only the most probable distribution of the drawings from an urn, as described above, but it serves also as a norm for the distribution of relative frequencies obtained from some of the simplest sampling operations in applied statistics. For example, the geneticist may regard the Bernoulli distribution (1) as the theoretical distribution of the relative frequencies m/s of green peas which he would obtain

[2] See references on pp. 173–77.

among random samples each consisting of a yield of s peas. The biologist may regard (1) as the theoretical distribution of the relative frequencies of male births in random samples of s births. The actuary may regard (1) as the theoretical distribution of yearly death-rates in samples of s men of equal ages, say of age 30, drawn from a carefully described class of men. In this case we specify that the samples shall be taken from a carefully described class of men because the assumptions involved in the urn schemata underlying a Bernoulli distribution do not permit a careless selection of data. Thus, it would not be in accord with the assumptions to take some of the samples from a group of teachers with a relatively low rate of mortality and others from a group of anthracite coal miners with a relatively high rate of mortality.

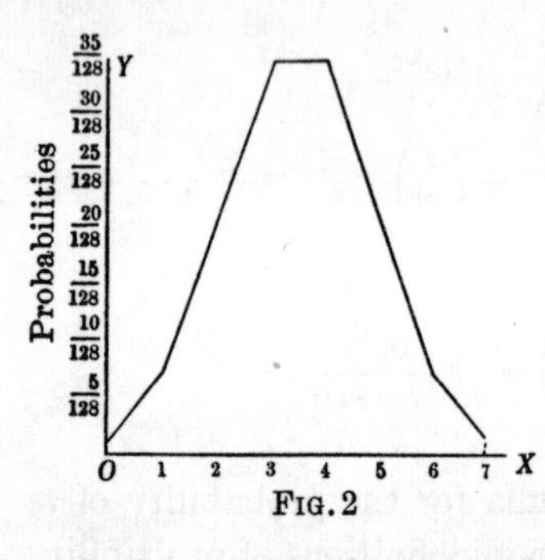

FIG. 2

The fact stated at the beginning of this section that we are concerned with repeating the process of drawing from the same population is intended to imply that the same set of circumstances essential to drawing a random sample shall exist throughout the whole series of drawings.

The expression "simple sampling" is sometimes applied to drawing a random sample when the conditions for repetition just described are fulfilled. In other words, simple sampling implies that we may assume the underlying probability p of formula (1) remains constant from sample to sample, and that the drawings are mutually independent in the sense that the results of drawings do not depend in any significant manner on what has happened in previous drawings.

In Figure 2 the ordinates at $x=0, 1, 2, \ldots, 7$ show the values of terms of (1) for $p=q=1/2$, $s=7$. To find the "most probable" or modal number of successes m' in s trials, we seek the value of $m=m'$ which gives a maximum term of (1). To find this value of m, we write the ratios of the general term of (1) to the preceding and the succeeding terms. The first ratio will be equal to or greater than unity when

$$\frac{s-m+1}{m}\frac{p}{q} \geqq 1 \text{ or } m \leqq ps+p .$$

In the same way, the second ratio will be equal to or greater than unity when

$$\frac{m+1}{s-m}\frac{q}{p} \geqq 1 \text{ or } m \geqq ps-q .$$

We have, thus, the integer $m=m'$ which gives the modal value determined by the inequalities,

$$ps-q \leqq m' \leqq ps+p .$$

We may say therefore that, neglecting a proper fraction, ps is the most probable or modal number of successes. When $ps-q$ and $ps+p$ are integers, there occur two equal terms in (1) each of which is larger than any other term of the series. For example, note the equality of the first and second terms of the expansion $(5/6+1/6)^5$.

10. Mathematical expectation and standard deviation of the number of successes. Let $\bar{m}$ be the mathematical expectation of the number of successes, or what is the same thing, the arithmetic mean number of successes in s trials under the law of repeated trials as defined by formula (1) on page 23. We shall now prove that $\bar{m}=ps$.

By definition (§ 6),

$$(2)\quad \begin{cases} \bar{m}=\displaystyle\sum_{m=0}^{m=s}\frac{s!}{m!\,(s-m)!}\,p^m\,q^{s-m}m \\ \quad =\displaystyle\sum_{m=1}^{m=s}\frac{s!}{(m-1)!\,(s-m)!}\,p^m\,q^{s-m} \end{cases}$$

$$(3)\qquad =sp\sum_{m=1}^{m=s}\frac{(s-1)!}{(m-1)!\,(s-m)!}\,p^{m-1}\,q^{s-m}=sp\ ,$$

since

$$\sum_{m=1}^{m=s}\binom{s-1}{m-1}p^{m-1}\,q^{s-m}=(p+q)^{s-1}=1\ .$$

Let $d=m-sp$ be the *discrepancy* of the number of successes from the mathematical expectation, and let σ^2 be the mathematical expectation of the square of the discrepancy. By definition,

$$(4)\quad \begin{cases} \sigma^2=\displaystyle\sum_{m=0}^{m=s}\frac{s!}{m!\,(s-m)!}\,p^m\,q^{s-m}(m-sp)^2 \\ \quad =\displaystyle\sum_{m=0}^{m=s}\frac{s!}{m!\,(s-m)!}\,p^m\,q^{s-m}(m^2-2msp+s^2p^2)\ . \end{cases}$$

We shall now prove that $\sigma^2 = spq$. To do this, we write $m^2 = m + m(m-1)$ and obtain for the first term of (4) the value

$$(5)\quad \begin{cases} \displaystyle\sum_{m=0}^{m=s} \frac{s!\,m}{m!\,(s-m)!}\, p^m q^{s-m} \\ \displaystyle\qquad + s(s-1)p^2 \sum_{m=2}^{m=s} \frac{(s-2)!}{(m-2)!\,(s-m)!}\, p^{m-2} q^{s-m} \\ \qquad\qquad = sp + s(s-1)p^2 . \end{cases}$$

From (2), (3), (4), and (5), we have

$$(6)\quad \begin{cases} \sigma^2 = sp + s(s-1)p^2 - 2s^2p^2 + s^2p^2 \\ \quad\; = sp(1-p) = spq . \end{cases}$$

The measure of dispersion σ is often called the *standard deviation* of the frequency of successes in the population.

Next, we define $d/s = (m/s) - p$ as the *relative discrepancy*, for it is the difference between the probability of success and the relative frequency of success. The mean square of the relative discrepancy is the second member of equation (4) divided by s^2. It is clearly equal to the mean square σ^2 of the discrepancy divided by s^2, which gives

$$(7)\quad \frac{pq}{s} .$$

The theoretical value of the standard deviation of the relative frequency of successes is then $(pq/s)^{1/2}$.

11. **Theorem of Bernoulli.** The theorem of Bernoulli deals with the fundamental problem of the approach of the relative frequency m/s of success in s trials to the

underlying constant probability p as s increases. The theorem may be stated as follows:

In a set of s trials in which the chance of a success in each trial is a constant p, the probability P of the relative discrepancy $(m/s)-p$ being numerically as large as any assigned positive number ϵ will approach zero as a limit as the number of trials s increases indefinitely, and the probability, $Q=1-P$, of this relative discrepancy being less than ϵ approaches 1 *or certainty.*

This theorem is sometimes called the law of large numbers. The theorem has been very commonly regarded as the basic theorem of mathematical statistics. But with the definition of probability (p. 8) as the limit of the relative frequency, this theorem is an immediate consequence of the definition. While it adds to the definition something about the manner of approach to the limit the theorem is in some respects not so strong as the corresponding assumption in the definition.

With a definition of probability other than the limit definition, the theorem may not follow so readily. It has been regarded as fundamental because of its bearing on the use of the relative frequency m/s (s large) as if it were a close approximation to the probability p. Assuming for the present that we have any definition of the probability p of success in one trial from which we reach the law of repeated trials given in the binomial expansion (1), we may prove the Bernoulli theorem by the use of the Bienaymé-Tchebycheff criterion.

To derive this criterion, consider a statistical variable x which takes mutually exclusive values $x_1, x_2, \ldots, x_n$ with probabilities $p_1, p_2, \ldots, p_n$, respectively, where

$$p_1+p_2+\cdots+p_n=1 .$$

Let a be any given number from which we wish to measure deviations of the x's. A specially important case is that in which a is a mean or expected value of x, although a need not be thus restricted. For the expected mean-square deviation from a, we may write

$$\sigma^2 = p_1 d_1^2 + p_2 d_2^2 + \cdots + p_n d_n^2,$$

where $d_i = x_i - a$.

Let $d', d'', \ldots,$ be those deviations $x_i - a$ which are at least as large numerically as an assigned multiple $\epsilon = \lambda\sigma (\lambda > 1)$ of the root-mean-square deviation σ from a, and let $p', p'', \ldots,$ be the corresponding probabilities. Then we have

$$\sigma^2 \geqq p' d'^2 + p'' d''^2 + \cdots. \tag{8}$$

Since $d', d'', \ldots,$ are each numerically equal to or greater than $\lambda\sigma$, we have from (8) that

$$\sigma^2 \geqq \lambda^2 \sigma^2 (p' + p'' + \cdots).$$

If we now let $P(\lambda\sigma)$ be the probability that a value of x taken at random from the "population" will differ from a numerically by as much as $\lambda\sigma$, then $P(\lambda\sigma) = p' + p'' + \cdots,$ and $\sigma^2 \geqq \lambda^2 \sigma^2 P(\lambda\sigma)$. Hence

$$P(\lambda\sigma) \leqq \frac{1}{\lambda^2}. \tag{9}$$

To illustrate numerically we may take a to be the arithmetic mean of the x's and say that the probability is not more than 1/25 that a variate taken at random will deviate from the arithmetic mean as much as five times the standard deviation.

A striking property of the Bienaymé-Tchebycheff criterion is its independence of the nature of the distribution of the given values.

In a slightly different form, we may state that the probability is greater than $1-1/\lambda^2$, that a variate taken at random will deviate less than $\lambda\sigma$ from the mathematical expectation. This theorem is ordinarily known as the inequality of Tchebycheff,[3] but the main ideas underlying the inequality were also developed by Bienaymé.[4]

We shall now turn our attention more directly to the theorem of Bernoulli. We seek the probability that the relative discrepancy $(m/s)-p$ will be numerically as large as an assigned positive number ϵ.

We may take $\epsilon=\lambda(pq/s)^{1/2}$, a multiple of the theoretical standard deviation $(pq/s)^{1/2}$ of the relative frequencies m/s. (See §10).

Let P be the probability that

$$\left|\frac{m}{s}-p\right| \geqq \lambda\left(\frac{pq}{s}\right)^{1/2},$$

then from the Bienaymé-Tchebycheff criterion (9), we have $P\leqq 1/\lambda^2$.

Since

$$\frac{1}{\lambda}=\frac{1}{\epsilon}\left(\frac{pq}{s}\right)^{1/2}, \text{ we have } P\leqq\frac{pq}{s\epsilon^2}.$$

For any assigned ϵ, we may by increasing s make P small at pleasure. That is, the probability P that the relative frequency m/s will differ from the probability p by at least as much as an assigned number, however small, tends toward zero as the number of cases s is indefinitely increased.

For example, if we are concerned with the probability P that $|(m/s)-p| \geqq .001$, we see that $P \leqq 1{,}000{,}000pq/s$. If the number of trials s is not very large, this inequality would ordinarily put no important restriction on P. But as s increases indefinitely, $1{,}000{,}000pq$ remains constant, and $1{,}000{,}000pq/s$ approaches zero. Again, the probability $Q=1-P$ that $|(m/s)-p|$ is less than ϵ satisfies the condition

$$Q>1-\frac{pq}{s\epsilon^2}. \tag{10}$$

From (10) we see that with any constant pq/ϵ^2, the probability Q becomes arbitrarily near 1 or certainty as s increases indefinitely. Hence the theorem is established for any definition of probability from which we derive (1) as the law of repeated trials.

It seems that the statement of the theorem concerning the probable approach of relative frequencies to the underlying probability may appear simpler and more elegant by the use of the concept of asymptotic certainty introduced by E. L. Dodd in a recent paper.[5] According to this concept, we may say it is asymptotically certain that m/s will approach p as a limit as s increases indefinitely.

12. **The De Moivre-Laplace Theorem.**[6] The De Moivre-Laplace theorem deals with the probability that the number of successes m in a set of s trials will fall within a certain conveniently assigned discrepancy d from the mathematical expectation sp. By the inequality of Tchebycheff (p. 30) a lower limit to the value of this probability has been given. We shall now proceed to con-

sider the problem of finding at least the approximate value of the probability. This problem would, in the simplest cases, involve merely the evaluation and addition of certain terms of the expansion (1). But this procedure would, in general, be impracticable when s is large and d even fairly large. To visualize the problem we represent the terms of (1) by ordinates y_x at unit intervals where x marks deviations of m from the mathematical expectation of successes ps as an origin. Then we have

$$y_x = \frac{s!}{(ps+x)!\,(qs-x)!}\, p^{ps+x}\, q^{qs-x}\,. \tag{11}$$

The probability that the number of successes will lie within the interval $ps-d$ and $ps+d$, inclusive of end values, is then the sum of the ordinates

$$y_{-d}+y_{-(d-1)}+\cdots\cdots+y_0+y_1+\cdots\cdots+y_d=\sum_{-d}^{+d} y_x\,. \tag{12}$$

As the number of y's in this sum is likely to be large, some convenient method of finding the approximate value of the sum will be found useful. In attacking this problem, we shall first of all replace the factorials in (11) approximately by the first term of Stirling's formula for the representation of large factorials.

This formula[7] states that

$$n! = n^n e^{-n}(2\pi n)^{1/2}\left(1+\frac{1}{12n}+\frac{1}{288n^2}+\cdots\right). \tag{13}$$

To form an idea of the degree of approximation obtained by using only the first term of this formula, we may say that in replacing $n!$ by $n^n e^{-n}(2\pi n)^{1/2}$ we obtain a result equal to the true value divided by a number between 1 and $1+1/10n$. The use of this first term is thus a sufficiently close approximation for many purposes if n is fairly large. The substitution by the use of Stirling's formula for factorials in (11) gives, after some algebraic simplification,

$$(14) \qquad y_x = \frac{1}{(2\pi spq)^{1/2}} \left(1+\frac{x}{ps}\right)^{-ps-x-\frac{1}{2}} \left(1-\frac{x}{qs}\right)^{-qs+x-\frac{1}{2}}$$

approximately.

To explain further our conditions of approximation to (11), we naturally compare any individual discrepancy x from the mathematical expectation ps with the standard deviation $\sigma=(spq)^{1/2}$. We should note in this connection that σ is of order $s^{1/2}$ if neither p nor q is extremely small. This fact suggests the propriety of assuming that s is so large that x/s shall remain negligibly small, but that $x/s^{1/2}$ may take finite values such as interest us most when we are making comparisons of a discrepancy with the standard deviation. It is important to bear in mind that we are for the present dealing with a particular kind of approximation.

Under the prescribed conditions of approximation, we shall now examine (14) with a view to obtaining a more convenient form for y_x. For this purpose, we may write

$$(15) \qquad \begin{cases} \log\left(1+\dfrac{x}{sp}\right)^{-sp-x-\frac{1}{2}} \\ \qquad = -(sp+x+\frac{1}{2})\left[\dfrac{x}{sp}-\dfrac{x^2}{2s^2p^2}+\dfrac{x^3}{s^3}\,\phi(x)\right], \end{cases}$$

$$(16) \qquad \begin{cases} \log\left(1-\dfrac{x}{sq}\right)^{-sq+x-\frac{1}{2}} \\ \qquad = -(sq-x+\frac{1}{2})\left[-\dfrac{x}{sq}-\dfrac{x^2}{2s^2q^2}-\dfrac{x^3}{s^3}\,\phi_1(x)\right], \end{cases}$$

where $\phi(x)$ and $\phi_1(x)$ are finite because each of them represents the sum of a convergent power series when x/s is small at pleasure. From (14), (15), and (16),

$$\log y_x(2\pi spq)^{1/2} = \frac{(p-q)x}{2spq} - \frac{x^2}{2spq} + \frac{x^2}{s^2}\phi_2(x)$$

$$= -\frac{x^2}{2spq} + \frac{x}{s}\phi_3(x),$$

where $\phi_3(x)$ is clearly finite.

Now if s is so large that $(x/s)\,\phi_3(x)$ becomes small, we have

$$y_x = \frac{1}{(2\pi spq)^{1/2}} e^{-\frac{x^2}{2spq}}$$

as an approximation to y_x in (11).

As a first approximation to the sum of the ordinates in (12), we then write the integral

$$\text{(17)} \qquad \frac{1}{(2\pi spq)^{1/2}} \int_{-d}^{+d} e^{-\frac{x^2}{2spq}} dx .$$

This integral is commonly known as the probability integral. The ordinates of the bell-shaped curve (Fig. 3) represent the values of the function

$$y = \frac{1}{\sigma(2\pi)^{1/2}} e^{-\frac{x^2}{2\sigma^2}},$$

where $\sigma^2 = spq$. This curve is the normal frequency curve and will be further considered in Chapter III.

We may increase slightly the accuracy of our approximation by taking account of the fact that we have one more ordinate in (12) than intervals of area. We may therefore appropriately add an ordinate at $x=d$ to the value given in (17), and obtain

$$(18) \qquad \frac{1}{(2\pi spq)^{1/2}}\int_{-d}^{d} e^{-\frac{x^2}{2spq}}\,dx + \frac{1}{(2\pi spq)^{1/2}}\, e^{-\frac{d^2}{2spq}},$$

for the probability that the discrepancy is between $-d$ and d inclusive of end points.

Another method of taking account of the extra ordinate is to extend the limits of integration in (17) by

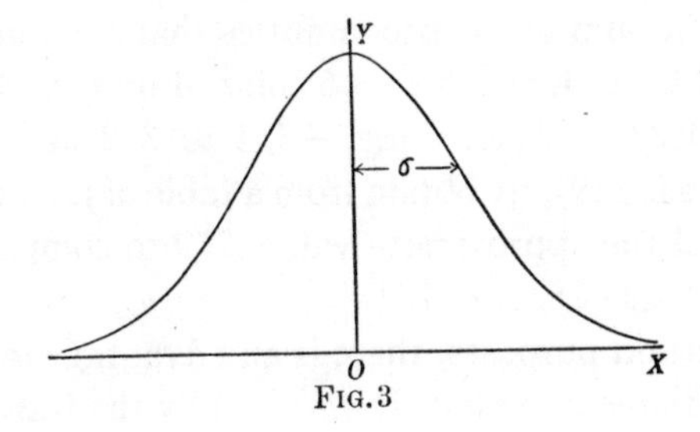

FIG. 3

one-half the unit at both the upper and lower limits. That is, we write

$$(19) \qquad \frac{1}{(2\pi spq)^{1/2}}\int_{-d-\frac{1}{2}}^{d+\frac{1}{2}} e^{-\frac{x^2}{2spq}}\,dx$$

in place of (17).

We may now state the De Moivre-Laplace theorem: *Given a constant probability p of success in each of s trials where s is a large number, the probability that the discrepancy $m-sp$ of the number m of successes from the mathematical expectation will not exceed numerically a given positive number d is given to a first approximation by* (17) *and to closer approximations by* (18) *and* (19).

Although formulas (17), (18), and (19) assume s large, it is interesting to experiment by applying these formulas to cases in which s is not large. For example, consider the problem of tossing six coins. The most probable number of heads is 3, and the probability of a discrepancy equal to or less than 1 is given exactly by

$$\left(\frac{6!}{3!\,3!}+\frac{6!}{4!\,2!}+\frac{6!}{2!\,4!}\right)\frac{1}{64}=\frac{25}{32},$$

which is the sum of the probabilities that the number of heads will be 2, 3, or 4 for $s=6$ coins. But $spq=1.5$, and $(spq)^{1/2}=1.225$. Then using $-3/2$ to $3/2$ as limits of integration in (19), we obtain from a table of the probability integral the approximate value .779 to compare with the exact value $25/32=.781$.

For certain purposes, there is an advantage in changing the variable x to t in (17) and (18) by the transformation

$$\frac{x}{(spq)^{1/2}}=t, \qquad \frac{d}{(spq)^{1/2}}=\delta.$$

Then in place of (17) we have

$$P_\delta=\left(\frac{2}{\pi}\right)^{1/2}\int_0^\delta e^{-\frac{t^2}{2}}\,dt, \tag{20}$$

and in place of (18) we have

$$P_\delta=\left(\frac{2}{\pi}\right)^{1/2}\int_0^\delta e^{-\frac{t^2}{2}}\,dt+\frac{e^{-\frac{\delta^2}{2}}}{(2\pi spq)^{1/2}}. \tag{21}$$

To give a general notion of the magnitude of the probabilities, we shall now list a few values of P_δ in (20) corresponding to assigned values of δ. Thus,

δ ...	0	.6745	1	2	3	4
P_δ ...	0	.5	.68269	.95450	.99730	.99994

Extensive tables giving values of the probability integral and of the ordinates of the probability curve are readily available. For example, the Glover *Tables of Applied Mathematics*[8] give $P_\delta/2$ for the argument $\delta=x/\sigma$. Sheppard's table[9] gives $(1+P_\delta)/2$ for the argument $\delta=x/\sigma$.

We may now state the De Moivre-Laplace theorem in another form by saying that the values of P_δ in (20) and (21) give approximations to the probability that $|m-sp|<\delta(spq)^{1/2}$ for an assigned positive value of δ.

In still another slightly different form involving relative frequencies, we may state that the values of P_δ in (20) and (21) give approximations to the probability that the absolute value of the relative discrepancy satisfies the inequality

$$\left|\frac{m}{s}-p\right|<\delta\left(\frac{pq}{s}\right)^{1/2} \tag{22}$$

for every assigned positive value of δ.

In order to gain a fuller insight into the significance of the De Moivre-Laplace theorem we may draw the following conclusions from (20): (*a*) Assuming as is suggested by (20) that a δ exists corresponding to every assigned probability P_δ, we find from $d=\delta(spq)^{1/2}$ that the bounds $-d$ to $+d$ increase in proportion to $s^{1/2}$ as s is increased (*b*) From (20) and (22) it follows that for assigned probabilities P_δ the bounds of discrepancy of the relative frequency m/s from p vary inversely as $s^{1/2}$.

To illustrate the use of the De Moivre-Laplace theorem, we take an example from the third edition of the *American Men of Science* by Cattell and Brimhall (p. 804). A group of scientific men reported 1,705 sons and 1,527 daughters. The examination of these numbers brings up the following fundamental questions of simple sampling. Do these data conform to the hypothesis that 1/2 is the probability that a child to be born will be a boy? That is, can the deviations be reasonably regarded as fluctuations in simple sampling under this hypothesis? In another form, what is the probability in throwing 3,232 coins that the number of heads will differ from $(3{,}232/2)=1{,}616$ by as much, as or more than, $1{,}705-1{,}616=89$?

In this case,

$$s=3{,}232\ , \qquad (pqs)^{1/2}=28.425\ ,$$

$$d=1{,}705-1{,}616=89\ , \qquad \frac{d}{(pqs)^{1/2}}=3.131\ .$$

Referring to a table of the normal probability integral, we find from (20) that $P=.9983$. Hence, the probability that we will obtain a deviation more than 89 on either side of 1,616 in a single trial is approximately $1-.9983=.0017$.

13. **The quartile deviation.** The discrepancy d which corresponds to the probability $P=1/2$ in (20) is sometimes called the *quartile deviation*, or the *probable error* of m as an approximation to sp.

By the use of a table of the probability integral, it is found from (20) that $d=.6745\ (spq)^{1/2}$ approximately when $P=1/2$, and thus $.6745(spq)^{1/2}$ is the quartile deviation of the number of successes from the expectation sp.

14. **The law of small probabilities. The Poisson exponential function.** The De Moivre-Laplace theorem does not ordinarily give a good approximation to the terms of the binomial $(p+q)^s$ if p or q is small. If s is large but sp or sq is small in relation to s, we may give a useful representation of terms of the binomial expansion $(p+q)^s$ by means of the Poisson exponential function. Statistical examples of this situation are what may be called rare events and may easily be given: The number born blind per year in a city of 100,000, or the number dying per year of a minor disease.

Poisson[10] had already as early as 1837 given the function involved in the treatment of the problem. Bortkiewicz[11] took up the problem in connection with a long series of observations of events which occur rarely. For example, one well-known series he gave was the frequency distribution of the number of men killed per army corps per year in the Prussian army from the kicks of horses. The frequency distribution of the number of deaths per army corps per year was:

Deaths	0	1	2	3	4
Frequency	109	65	22	3	1

He called the law of frequency involved the "law of small numbers," and this name continues to be used although it does not seem very appropriate. The expression "law of small probabilities" seems to give a more accurate description. Assume, then, that the probability p is small and that $q=1-p$ is nearly unity. That is, p is the probability of the occurrence of the rare event in question in a single trial.

We then seek a convenient expression approximately equal to

$$P_m = \frac{s!}{m!\,n!}\, p^m q^n\,,$$

the probability of m occurrences and n non-occurrences in $m+n=s$ trials.

Replacing $s!$ and $n!$ by means of Stirling's formula we obtain

$$P_m = \frac{n^m}{(1-m/s)^{s+\frac{1}{2}} m!}\, e^{-m} p^m q^n\,.$$

With large values of s and relatively small values of m, $(1-m/s)^{s+1/2}$ differs relatively little from $(1-m/s),^s$ and this in turn differs relatively little from e^{-m}. Furthermore, $q^n=(1-p)^n$ differs very little from e^{-np} since, on the one hand,

$$(1-p)^n = 1-np+\frac{n(n-1)}{2}\, p^2 - \cdots\,;$$

and, on the other,

$$e^{-np} = 1-np+\frac{n^2p^2}{2} - \cdots\,.$$

Introducing these approximations by substituting e^{-m} for $(1-m/s)^{s+1/2}$, and e^{-np} for q^n, we have

$$P_m = \frac{n^m}{m!}\, p^m e^{-np}\,.$$

For rare events, of small probability p, np differs very little from $sp=\lambda$. Hence, we write

$$P_m=\frac{\lambda^m e^{-\lambda}}{m!} \tag{23}$$

for the approximate probability of m occurrences of the rare event. Then the terms of the series

$$e^{-\lambda}\left(1+\lambda+\frac{\lambda^2}{2!}+\frac{\lambda^3}{3!}+\cdots\right)$$

give the approximate probabilities of exactly 0, 1, 2, ... , occurrences of the rare event in question, and the sum of series

$$e^{-\lambda}\left(1+\lambda+\frac{\lambda^2}{2!}+\frac{\lambda^3}{3!}+\cdots+\frac{\lambda^m}{m!}\right) \tag{24}$$

gives the probability that the rare event F will happen either 0, 1, 2, ... , or m times in s trials.

Although we have assumed in deriving the Poisson exponential function $\lambda^m e^{-\lambda}/m!$ that m is small in comparison with s, we may obtain certain simple and interesting results for the mathematical expectation and standard deviation of the distribution given by the Poisson exponential when m takes all integral values from $m=0$ to $m=s$. Thus, when $m=s$ in (24), we clearly have

$$e^{-\lambda}\left(1+\lambda+\frac{\lambda^2}{2!}+\frac{\lambda^3}{3!}+\cdots+\frac{\lambda^s}{s!}\right)=1 \tag{25}$$

approximately if s is large.

Since the successive terms in (25) give approximately the probabilities of 0, 1, 2, ... , s occurrences of the rare event, the mathematical expectation $\sum mP_m$ of the number of such occurrences is

$$e^{-\lambda}\left(0+\lambda+\lambda^2+\frac{\lambda^3}{2!}+\cdots+\frac{\lambda^s}{(s-1)!}\right)=$$

$$\lambda e^{-\lambda}\left(1+\lambda+\frac{\lambda^2}{2!}+\cdots+\frac{\lambda^{s-1}}{(s-1)!}\right)=\lambda$$

approximately when s is large.

Similarly, the second moment μ_2' about the origin is

$$\mu_2'=e^{-\lambda}\left[\lambda+2\lambda^2+\frac{3\lambda^3}{2!}+\cdots+\frac{s\lambda^s}{(s-1)!}\right],$$

and the second moment about the mathematical expectation is

$$(26)\quad\left\{\begin{aligned}\mu_2&=\mu_2'-\lambda^2=\lambda e^{-\lambda}\left[1+2\lambda+\frac{3\lambda^2}{2!}+\cdots+\frac{s\lambda^{s-1}}{(s-1)!}\right]-\lambda^2\\&=\lambda e^{-\lambda}\left[1+\lambda+\frac{\lambda^2}{2!}+\frac{\lambda^{s-1}}{(s-1)!}\right]\\&\qquad+\lambda^2e^{-\lambda}\left[1+\lambda+\frac{\lambda^2}{2!}+\cdots+\frac{\lambda^{s-2}}{(s-2)!}\right]-\lambda^2\\&=\lambda+\lambda^2-\lambda^2\text{ nearly }=sp\,,\end{aligned}\right.$$

an approximation to spq since q differs but little from 1.

Tables of the Poisson exponential limit $e^{-\lambda}\lambda^x/x!$ are given in *Tables for Statisticians and Biometricians* (pp. 113–24), and in *Biometrika*, Volume 10 (1914), pages 25–35. The values of $e^{-\lambda}\lambda^x/x!$ are tabulated to six places of decimals for λ varying from .1 to 15 by intervals of one-tenth and for x varying from 0 to 37.

A general notion of the values of the function for certain values of λ may be obtained from Figure 4 where the ordinates at 0, 1, 2, ... , show the values of the function for $\lambda = .5$, 1, 2, and 5.

Miss Whittaker has prepared special tables (*Tables for Statisticians and Biometricians*, pp. 122–24) which facilitate the comparison of results from the Poisson exponential with those from the De Moivre-Laplace theory in dealing with the sampling fluctuations of small frequencies. The question naturally arises as to the value of p below which we should prefer to use the Poisson exponential in dealing with the probability of a discrepancy less than an assigned number in place of the results of the De Moivre-Laplace theory. While there is no exact answer to this question, there seems to be good reason for certain purposes in restricting the application of the De Moivre-Laplace results to cases where the probability is perhaps not less than $p = .03$.

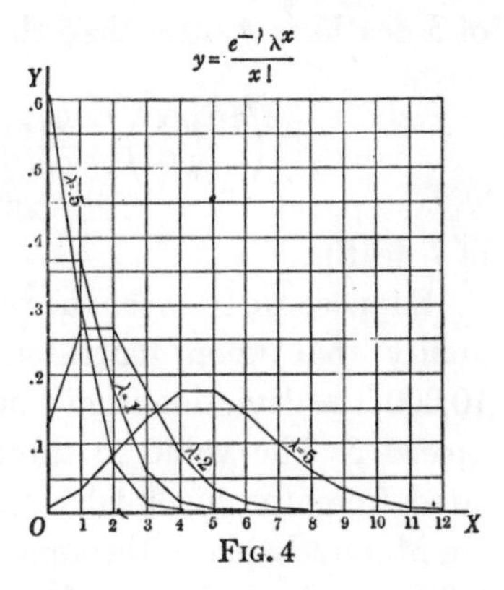

FIG. 4

To illustrate by a concrete situation in which p is small, consider a case of 6 observed deaths from pneumonia in an exposure of 10,000 lives of a well-defined class aged 30 to 31. It is fairly obvious, on the one hand, that the possible variations below 6 are restricted to 6, whereas there is no corresponding restriction above 6. On the other hand, if we take $(6/10{,}000) \doteq 3/5{,}000$ as the probability of death from pneumonia within a year of a person aged 30, it is more likely that we shall experience 5 deaths than 7 deaths among the 10,000 exposed; for the probability

$$\binom{10,000}{5}\left(\frac{4,997}{5,000}\right)^{9,995}\left(\frac{3}{5,000}\right)^{5}$$

of 5 deaths is greater than the probability

$$\binom{10,000}{7}\left(\frac{4,997}{5,000}\right)^{9,993}\left(\frac{3}{5,000}\right)^{7}$$

of 7 deaths.

Suppose we now set the problem of finding the probability that upon repetition with another sample of 10,000, the deviation from 6 deaths on either side will not exceed 3. The value to three significant figures calculated from the binomial expansion is .854. To use the De Moivre-Laplace theorem, we simply make $d=3$ in (19), and obtain from tables of probability functions the value $P_3=.847$.

We should thus expect from the De Moivre-Laplace theorem a discrepancy either in defect more than 3 or in excess more than 3 in $100-84.7=15.3$ per cent of the cases, and from the sum of the binomial terms we should expect such a discrepancy in $100-85.4=14.6$ per cent of the cases.

Turning next to tables of the Poisson exponential, page 122 of *Tables for Statisticians and Biometricians*, we find that in 6.197 per cent of cases there will be a discrepancy in defect more than 3 and in 8.392 per cent of cases there will be a discrepancy in excess more than 3.

The sum of 6.197 and 8.392 per cent is 14.589 per cent. This result differs very little for purposes of dealing with sampling errors from the 15.3 per cent given by the De Moivre-Laplace formula, but it is a closer approximation to the correct value and has the advantage of showing separately the percentage of cases in excess more than the assigned amount and the percentage in defect more than the same amount.

CHAPTER III

FREQUENCY FUNCTIONS OF ONE VARIABLE

15. Introduction. In Chapter I we have discussed very briefly three different methods of describing frequency distributions of one variable—the purely graphic method, the method of averages and measures of dispersion, and the method of theoretical frequency functions or curves. The weakness and inadequacy of the purely graphic method lies in the fact that it fails to give a numerical description of the distribution. While the method of averages and measures of dispersion gives a numerical description in the form of a summary characterization which is likely to be useful for many statistical purposes, particularly for purposes of comparison, the method is inadequate for some purposes because (1) it does not give a characterization of the distribution in the neighborhood of each point x or in each small interval x to $x+dx$ of the variable, (2) it does not give a functional relation between the values of the variable x and the corresponding frequencies.

To give a description of the distribution at each small interval x to $x+dx$ and to give a functional relation between the variable x and the frequency or probability we require a third method, which may be described as the "analytical method of describing frequency distributions." This method uses theoretical frequency functions. That is, in this method of description we attempt to characterize the given observed frequency distribution by appealing to underlying probabilities, and we seek a frequency function $y=F(x)$ such that $F(x)dx$ gives to within infinitesimals of higher order the probability that a variate x' taken at random falls in the interval x to $x+dx$.

Although the great bulk of frequency distributions which occur so abundantly in practical statistics have certain important properties in common, nevertheless they vary sufficiently to present difficult problems in considering the properties of $F(x)$ which should be regarded as fundamental in the selection of an appropriate function to fit a given observed distribution.

The most prominent frequency function of practical statistics is the normal or so-called Gaussian function

$$y=\frac{1}{\sigma(2\pi)^{\frac{1}{2}}}e^{-\frac{x^2}{2\sigma^2}}, \tag{1}$$

where σ is the standard deviation (see Fig. 3, p. 35).

Although Gauss made such noteworthy contributions to error theory by the use of this function that his name is very commonly attached to the function, and to the corresponding curve, it is well known that Laplace made use of the exponential frequency function prior to Gauss by at least thirty years. It would thus appear that the name of Laplace might more appropriately be attached to the function than that of Gauss. But in a recent and very interesting historical note, Karl Pearson[6] finds that De Moivre as early as 1733 gave a treatment of the probability integral and of the normal frequency function. The work of De Moivre antedates the discussion of Laplace by nearly a half-century. Moreover, De Moivre's treatment is essentially our modern treatment. Hence it appears that the discovery of the normal function should be attributed to De Moivre, and that his name might be most appropriately attached to the function. It may well be recalled that we obtained this function (1) in the De Moivre-Laplace theory (p. 34). In (1) the origin is taken so that the x-co-ordinate of the centroid of area under the curve is zero. The approximate value of the centroid may be obtained from a large number of observed variates by finding their arithmetic mean. The σ is equal to the radius of gyration of the area under the curve with respect to the y-axis, and is obtained approximately from observed variates by finding their standard deviation. The probability or frequency function (1) has been derived from a great variety of hypotheses.[12] The difficulty is not one of deriving this function but rather one of establishing a high degree of probability that the hypotheses underlying the derivation are realized in relation to practical problems of statistics.

In the decade from 1890 to 1900, it became well established experimentally that the normal probability function is inadequate to represent many frequency distributions which arise in biological data. To meet the situation it was clearly desirable either to devise methods for characterizing the most conspicuous departures from the normal distributions or to develop generalized frequency curves. The description and characterization of these departures without the direct use of generalized frequency curves has been accomplished roughly by the introduction (see pp. 68–72) of measures of skewness and of peakedness (excess or kurtosis), but the rationale underlying such measures is surely to be sought most naturally in the properties of generalized frequency functions. In spite of the reasons which may thus be advanced for the study of generalized frequency curves, it is fairly obvious that, for the most part, the authors of the rather large number of recent elementary textbooks on the methods of statistical analysis seem to regard it as undesirable or impracticable to include in such books the theory of generalized frequency curves. The writer is inclined to agree with these authors in the view that the complications of a theory of generalized frequency curves would perhaps have carried them too far from their main purposes. Nevertheless, some results of this theory are important for elementary statistics in providing a set of norms for the description of actual frequency distributions. In order to avoid misunderstanding it should perhaps be said that it is not intended to imply that a formal mathematical representation of many numerical distributions is desirable, but rather that a certain amount of such representation of carefully selected distributions should be encouraged. A useful purpose will be served in this connection if we can make certain points of interest in the theory more accessible by means of the present monograph.

The problem of developing generalized frequency curves has been attacked from several different directions. Gram (1879), Thiele (1889), and Charlier (1905) in Scandinavian countries; Pearson (1895) and Edgeworth (1896) in England; and Fechner (1897) and Bruns (1897) in Germany have developed theories of generalized frequency curves from viewpoints which give very different degrees of prominence to the normal probability curve in the development of a more general theory. In the present monograph, special attention will be given to two systems of frequency curves—the Pearson system and the Gram-Charlier system.

16. **The Pearson system of generalized frequency curves.** Pearson's first memoir[13] dealing with generalized frequency curves appeared in 1895. In this paper he gave four types of frequency curves in addition to the normal curve, with three subtypes under his Type I and two subtypes under his Type III. He published a supplementary memoir[14] in 1901 which presented two further types. A second supplementary memoir[15] which was published in 1916 gave five additional types. Pearson's curves, which are widely different in general appearance, are so well known and so accessible that we shall take no time to comment on them as graduation curves for a great variety of frequency distributions, but we shall attempt to indicate the genesis of the curves with special reference to the methods by which they are grounded on or associated with underlying probabilities.

We shall consider a frequency function $y=F(x)$ of one variable where we assume that $F(x)dx$ differs at most by an infinitesimal of higher order from the probability that a variate x taken at random will fall into the interval x to $x+dx$. Pearson's types of curves $y=F(x)$ are obtained by integration of the differential equation

$$\frac{dy}{dx}=\frac{(x+a)y}{c_0+c_1x+c_2x^2}, \tag{2}$$

and by giving attention to the interval on x in which $y=F(x)$ is positive. The normal curve is given by the special case $c_1=c_2=0$. We may easily obtain a clear view of the genesis of the system of Pearson curves in relation to laws of probability by following the early steps in the development of equation (2). The development is started by representing the probabilities of successes in n trials given by the terms of the symmetric point binomial $(1/2+1/2)^n$ as ordinates of a frequency polygon. It is then easily proved that the slope dy/dx of any side of this polygon, at its midpoint, takes the form

$$\frac{dy}{dx}=-k^2(x+a)y\,, \tag{3}$$

where y is the ordinate at this point, and a and k are constants. By integration, we obtain the curve for which this differential equation is true at all points. The curve thus obtained is the normal curve (Pearson's Type VII).

The next step consists in dealing with the asymmetric point binomial $(p+q)^n$, $p\neq q$, in a manner analogous to that used in the case of the symmetric point binomial. This procedure gives the differential equation

$$\frac{dy}{dx}=\frac{(x+a)y}{c_0+c_1x}\,,$$

from which we obtain by integration the Pearson Type III curve

$$y=y_0\left(1+\frac{x}{a}\right)^{\gamma a}e^{-\gamma x}\,. \tag{4}$$

That is, with respect to the slope property, this curve stands in the same relation to the values given by the asymmetric binomial polygon as the normal curve does to values given by the symmetric binomial.

Thus far the underlying probability of success has

been assumed constant. The next step consists in taking up a probability problem in which the chance of success is not constant, but depends upon what has happened previously in a set of trials. Thus, the chance of getting r white balls from a bag containing np white and nq black balls in drawing s balls one at a time without replacements is given by

$$(5)\quad y_r=\binom{s}{r}\frac{(np)_r(nq)_{s-r}}{(n)_s}=\frac{(np)!\,(nq)!\,(n-s)!\,s!}{(np-r)!\,(nq-s+r)!\,n!\,r!\,(s-r)!}\,,$$

where $(n)_s$ means the number of permutations of n things taken s at a time and $\binom{s}{r}$ is the number of combinations of s things r at a time. This expression is a term of a hypergeometric series. By representing the terms of this series as ordinates of a frequency polygon, and finding the slope of a side of the frequency polygon, and proceeding in a manner analogous to that used in the case of the point binomial, we obtain a differential equation of the form given in (2). Thus, we make $r=0, 1, 2, \ldots, s$ and obtain the $s+1$ ordinates $y_0, y_1, y_2, \ldots, y_s$ at unit intervals. At the middle point of the side joining the tops of ordinates y_r and y_{r+1}, we have

$$(6)\qquad x=r+\tfrac{1}{2}\,,\qquad y=\tfrac{1}{2}(y_r+y_{r+1})\,,$$

and

$$(7)\quad \begin{cases} \dfrac{dy}{dx}=y_{r+1}-y_r=y_r\left[\dfrac{s-r}{r+1}\,\dfrac{np-r}{nq+r+1-s}-1\right] \\[2ex] \quad =y_r\,\dfrac{s+nps-nq-1-r(n+2)}{(r+1)(r+1+nq-s)}\,. \end{cases}$$

From $y=(y_r+y_{r+1})/2$, we have

$$(8)\quad \begin{cases} y=\frac{1}{2}y_r\left[\dfrac{(np-r)(s-r)}{(r+1)(r+1+nq-s)}+1\right] \\ \quad =\frac{1}{2}y_r\dfrac{nps+nq+1-s+r(nq+2-np-2s)+2r^2}{(r+1)(r+1+nq-s)}. \end{cases}$$

From (7) and (8), replacing r by $x-1/2$, we have

$$(9)\quad \frac{1}{y}\frac{dy}{dx}=\frac{2s+2nps-2nq-2-(2x-1)(n+2)}{nps+nq+1-s+(x-\frac{1}{2})(nq+2-np-2s)+2(x-\frac{1}{2})^2}.$$

From (9), we observe that the slope of the frequency polygon, at the middle point of any side, divided by the ordinate at that point is equal to a fraction whose numerator is a linear function of x and whose denominator is a quadratic function of x.

The differential equation (2) gives a general statement of this property. It is more general than (9) in that the constants of (9) are special values found from the law of probability involved in drawings from a limited supply without replacements. One of Pearson's generalizations therefore consists in admitting as frequency curves all those curves of which (2) is the differential equation without the limitations on the values of the constants involved in (9).

The questions involved in the integration of (2) and in the determination of parameters for actual distributions are so available in Elderton's *Frequency Curves and Correlation*, and elsewhere, that it seems undesirable to take the space necessary to deal with these questions here. The resulting types of equations and figures that indicate the general form of the curves for certain positive values of the parameters are listed below.

TYPE I (FIG. 5)

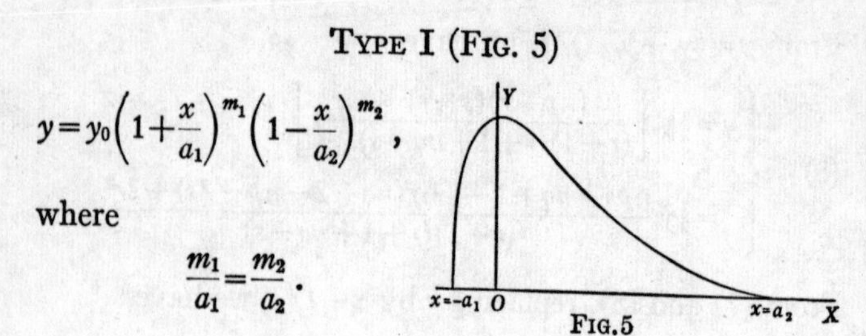

$$y=y_0\left(1+\frac{x}{a_1}\right)^{m_1}\left(1-\frac{x}{a_2}\right)^{m_2},$$

where

$$\frac{m_1}{a_1}=\frac{m_2}{a_2}.$$

FIG. 5

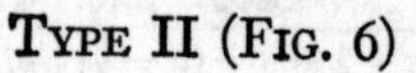

TYPE II (FIG. 6)

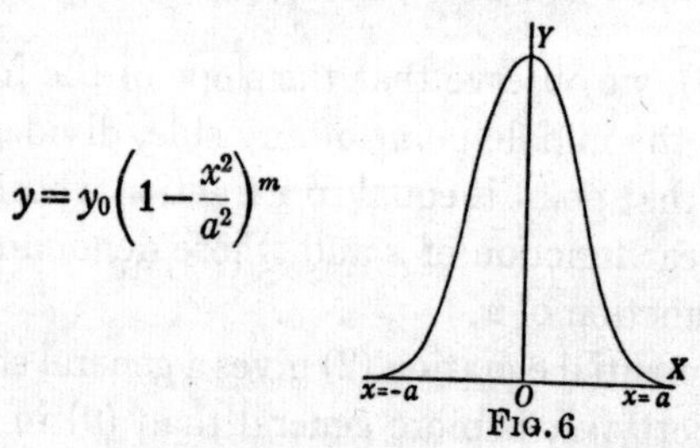

$$y=y_0\left(1-\frac{x^2}{a^2}\right)^m$$

FIG. 6

TYPE III (FIG. 7)

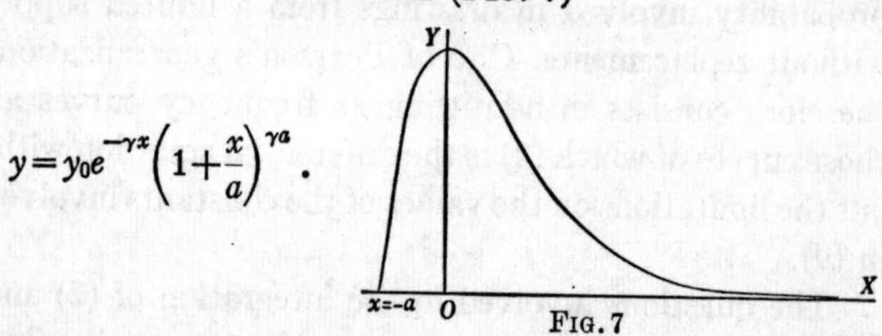

$$y=y_0e^{-\gamma x}\left(1+\frac{x}{a}\right)^{\gamma a}.$$

FIG. 7

TYPE IV

$$y=y_0\left(1+\frac{x^2}{a^2}\right)^{-m}e^{-\nu\arctan\frac{x}{a}}.$$

A skew curve of unlimited range at both ends, roughly described in general appearance as a slightly deformed normal curve (for the normal curve, see Fig. 3, p. 35).

TYPE V (FIG. 8)

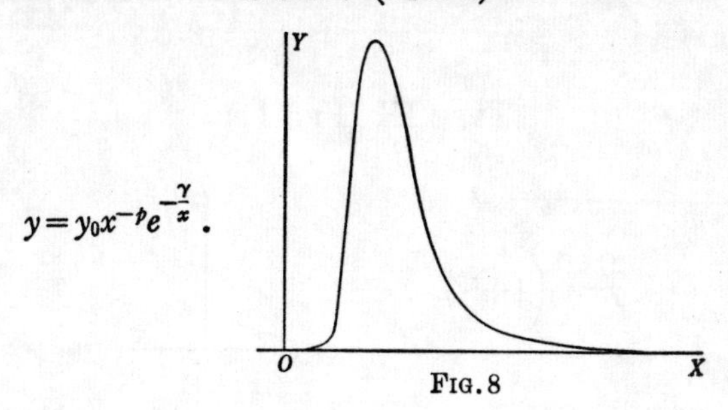

$$y = y_0 x^{-p} e^{-\frac{\gamma}{x}}.$$

FIG. 8

TYPE VI (FIG. 9)

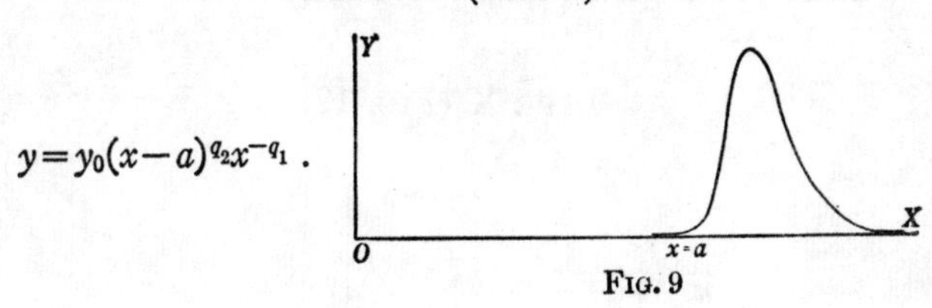

$$y = y_0 (x-a)^{q_2} x^{-q_1}.$$

FIG. 9

TYPE VII (FIG. 3, P. 35)

$$y = y_0 e^{-\frac{x^2}{2\sigma^2}}.$$

The normal frequency curve.

TYPE VIII (FIG. 10)

$$y = y_0 \left(1 + \frac{x}{a}\right)^{-m}.$$

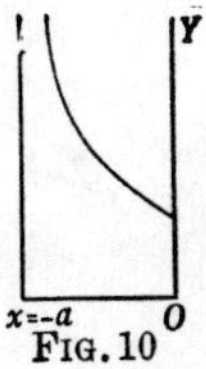

Y
x=-a
O

FIG. 10

This type degenerates into an equilateral hyperbola when $m=1$.

Type IX (Fig. 11)

$$y=y_0\left(1+\frac{x}{a}\right)^m .$$

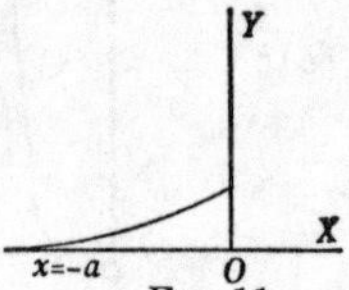

Fig. 11

This type degenerates into a straight line when $m=1$.

Type X (Fig. 12)

$$y=\frac{n}{\sigma}e^{\pm\frac{x}{\sigma}} .$$

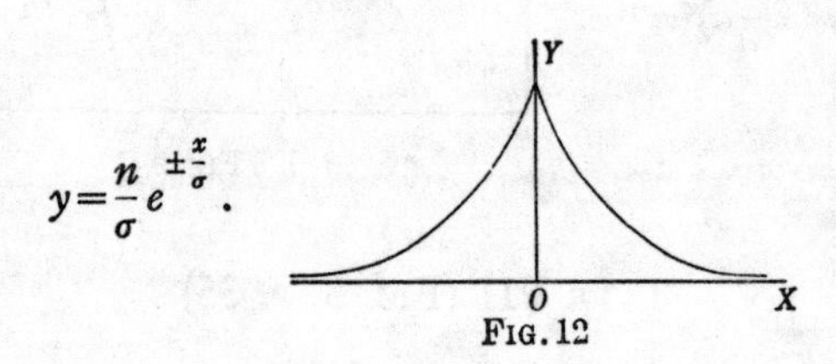

Fig. 12

This type is Laplace's first frequency curve while the normal curve is sometimes called his second frequency curve. The curve is shown for negative values of $\pm x/\sigma$.

Type XI (Fig. 13)

$$y=y_0x^{-m} .$$

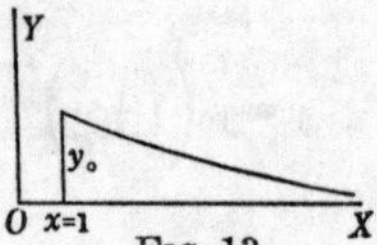

Fig. 13

TYPE XII (FIG. 14)

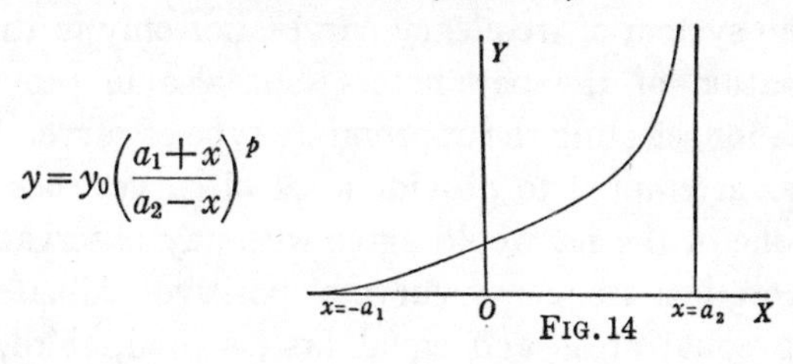

FIG. 14

The above figures should be regarded as roughly illustrating only in a meager way, for particular positive values of the parameters, the variety of shapes that are assumed by the Pearson type curves. For example, it is fairly obvious that Types I and II would be U-shaped when the exponents are negative, and that Type III would be J-shaped if γa were negative.

The idea of obtaining a suitable basis for frequency curves in the probabilities given by terms of a hypergeometric series is the main principle which supports the Pearson curves as probability or frequency curves, rather than as mere graduation curves. That is to say, these curves should have a wide range of applications as probability or frequency curves if the distribution of statistical material may be likened to distributions which arise under the law of probability represented by terms of a hypergeometric series, and if this law may be well expressed by determining a frequency function $y=F(x)$ from the slope of the frequency polygon of the hypergeometric series. In examining the source of the Pearson curves, the fact should not be overlooked that the normal probability curve can be derived from hypotheses containing much broader implications than are involved in a slope condition of the side of a symmetric binomial polygon.

The method of moments plays an essential rôle in the Pearson system of frequency curves, not only in the determination of the parameters, but also in providing criteria for selecting the appropriate type of curve. Pearson has attempted to provide a set of curves such that some one of the set would agree with any observational or theoretical frequency curve of positive ordinates by having equal areas and equal first, second, third, and fourth moments of area about a centroidal axis.

Let μ_m be the mth moment coefficient about a centroid vertical taken as the y-axis (cf. p. 19). That is, let

$$\mu_m = \int_{-\infty}^{\infty} x^m F(x)\,dx\,, \tag{10}$$

where $F(x)$ is the frequency function (see p. 13). Next, let

$$\beta_1 = \mu_3^2/\mu_2^3$$

and

$$\beta_2 = \mu_4/\mu_2^2\,.$$

Then it is Pearson's thesis that the conditions $\mu_0 = 1$, $\mu_1 = 0$, together with the equality of the numbers μ_2, β_1, and β_2, for the observed and theoretical curves lead to equations whose solutions give such values to the parameters of the frequency function that we almost invariably obtain excellency of fit by using the appropriate one of the curves of his system to fit the data, and that badness of fit can be traced, in general, to heterogeneity of data, or to the difficulty in the determination of moments from the data as in the case of J- and U-shaped curves.

Let us next examine the nature of the criteria by which to pass judgment on the type of curve to use in any numerical case. Obviously, the form which the integral $y=F(x)$ obtained from (2) takes depends on the nature of the zeros of the quadratic function in the denominator. An examination of the discriminant of this quadratic function leads to equalities and inequalities involving β_1 and β_2 which serve as criteria in the selection of the type of function to be used. A systematic procedure for applying these criteria has been thoroughly developed and published in convenient form in Pearson's *Tables for Statisticians and Biometricians* (1914), pages lx–lxx and 66–67; and in his paper in *The Philosophical Transactions*, A, Volume 216 (1916), pages 429–57. The relations between β_1 and β_2 may be conveniently represented by curves in the β_1-β_2 plane. Then the normal curve corresponds to the point $\beta_1=0$, $\beta_2=3$ in this plane. Type III is to be chosen when the point (β_1, β_2) is on the line $2\beta_2-3\beta_1-6=0$; and Type V, when (β_1, β_2) is on the cubic

$$\beta_1(\beta_2+3)^2=4(4\beta_2-3\beta_1)(2\beta_2-3\beta_1-6).$$

In considering subtypes under Type I, a biquadratic in β_1 and β_2 separates the area of J-shaped modeless curves from the area of limited range modal curves and the area of U-shaped curves.

Without going further into detail about criteria for the selection of the type of curve, we may summarize by saying that curves traced on the β_1 β_2-plane provide the means of selecting the Pearson type of frequency curve appropriate to the given distribution in so far as the necessary conditions expressed by relations between β_1 and β_2

turn out to be sufficient to determine a suitable type of curve.

The difficulties involved in the numerical computation of the parameters of the Pearson curves were rather clearly indicated in Pearson's original papers. The appropriate tables and forms for computations in fitting the curves to numerical distributions have been so available in various books as to facilitate greatly the applications to concrete data. Among such books and tables, special mention should be made of *Frequency Curves and Correlation* (1906), by W. P. Elderton, pages 5–105; *Tables for Statisticians and Biometricians* (1924), by Karl Pearson; and *Tables of Incomplete Gamma Functions* (1921), by the same author.

17. Generalized normal curves—Gram-Charlier series. Suppose some simple frequency function such as the normal function or the Poisson exponential function (p. 41) gives a rough approximation to a given frequency distribution and that we desire a more accurate analytic representation than would be given by the simple frequency function. In this situation, it seems natural to seek an analytical representation by means of the first few terms of a rapidly convergent series of which the first term, called the "generating function," is the simple frequency function which gives the rough approximation.

Prominent among the contributors to the method of the representation of frequency by a series may be named Gram,[16] Thiele,[17] Edgeworth,[18] Fechner,[19] Bruns,[20] Charlier,[21] and Romanovsky.[22]

Our consideration of series for the representation of frequency will be limited almost entirely to the Gram-Charlier generalizations of the normal frequency function and of the Poisson exponential function, by using these functions as generating functions. These two types of series may be written in the following forms:

TYPE A

$$F(x)=a_0\phi(x)+a_3\phi^{(3)}(x)+\cdots\cdots+a_n\phi^{(n)}(x)+\cdots\cdots, \tag{11}$$

where

$$\phi(x)=\frac{1}{\sigma(2\pi)^{1/2}}e^{-\frac{(x-b)^2}{2\sigma^2}},$$

and $\phi^{(n)}(x)$ is the nth derivative of $\phi(x)$ with respect to x.

TYPE B

$$F(x)=c_0\psi(x)+c_1\Delta\psi(x)+\cdots\cdots+c_n\Delta^n\psi(x)+\cdots\cdots, \tag{12}$$

where

$$\psi(x)=\frac{e^{-\lambda}\sin\pi x}{\pi}\left\{\frac{1}{x}-\frac{\lambda}{(x-1)1!}+\frac{\lambda^2}{(x-2)2!}-\cdots\cdots\right\}$$

$$=\frac{e^{-\lambda}\lambda^x}{x!},$$

which is the Poisson exponential for non-negative integral values of x, and where $\Delta\psi(x)$, $\Delta^2\psi(x)$, ... , denote the successive finite differences of $\psi(x)$ beginning with $\Delta\psi(x)=\psi(x)-\psi(x-1)$.

If Type A or Type B converges so rapidly that terms after the second or third may be neglected, it is fairly obvious that we have a simple analytic representation of the distribution.

The general appearance of the curves represented by two or three terms of Type A, for particular values of the coefficients, is shown in Figure 15 so as to facilitate com-

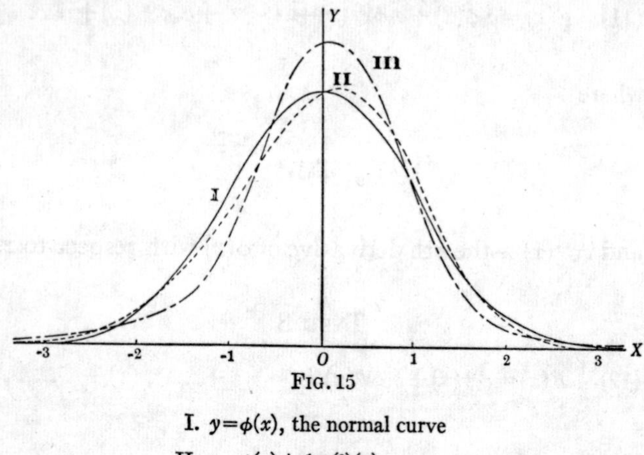

FIG. 15

I. $y=\phi(x)$, the normal curve

II. $y=\phi(x)+\frac{1}{15}\phi^{(3)}(x)$

III. $y=\phi(x)+\frac{1}{15}\phi^{(3)}(x)+\frac{1}{15}\phi^{(4)}(x)$

parison with the corresponding normal curve represented by the first term.

A general notion of the values of the function represented by the first term of Type B may be obtained for particular values of λ from Figure 4, page 43. When λ is taken equal to the arithmetic mean of the number of occurrences of the rare event in question, we shall find that $c_1=0$. We may then well inquire into the general appearance of the graph of the function

$$\psi(x)+c_2\Delta^2\psi(x)$$

for particular values of c_2 and λ. For $\lambda = 2$ and $c_2 = -.4$, see Figure 16, which shows also the corresponding $\psi(x)$.

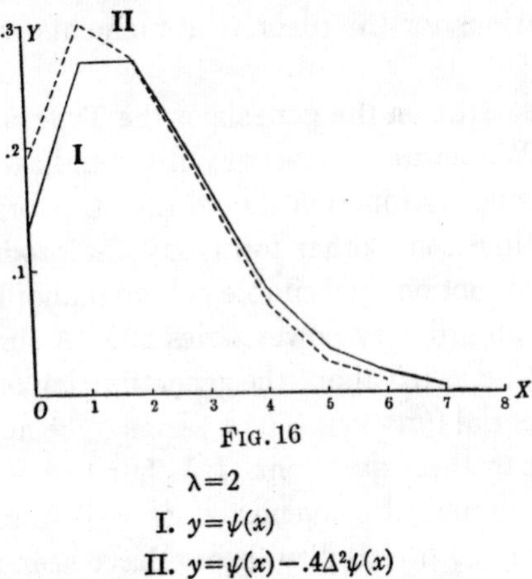

FIG. 16

$\lambda = 2$

I. $y = \psi(x)$

II. $y = \psi(x) - .4\Delta^2\psi(x)$

It should probably be emphasized that the usefulness of a series representation of a given frequency distribution depends largely upon the rapidity of convergence. In turn the rapidity with which the series converges depends much upon the degree of approach of the generating function to the given distribution.

Although it is known[23] that the Type A series is capable of converging to an arbitrary function $f(x)$ subject to certain conditions of continuity and vanishing at infinity, mere convergence is not sufficient for our problems. The representation of an actual frequency distribution requires, in general, such rapid convergence that only a few terms will be found necessary for the desired degree of approximation because (1) the amount of labor in computation soon becomes impracticable as the number of terms increases and (2) the probable errors of high-order moments involved in finding the parameters would generally be so large that the assumption that we may use moments of observations for the theoretical moments will become invalid.

18. **Remarks on the genesis of the Type A and Type B forms.** We naturally ask why a generalization of the normal frequency function should take the form of Type A rather than some other form, say the product of the generating function by a simple polynomial of low degree in x or by an ordinary power series in x. A similar question might be asked about the generalization of the Poisson exponential function. There seems to be no very simple answer to these questions. It is fair to say that algebraic and numerical convenience, as well as suggestions from underlying probability theory, have been significant factors in the selection of Type A and Type B. The algebraic and numerical convenience of Type A becomes fairly obvious by following Gram in determining the parameters. The suggestion of these forms in probability theory is closely associated with the development of the hypothesis of elementary errors (deviations) as given by Charlier.[21] A very readable discussion of the manner in which the Type A series arises in the probability theory of the distribution of a variate built up by the summation of a large number of independent elements is given in the recent book by Whittaker and Robinson on *The Calculus of Observations*, pages 168–74.

In the present monograph, we shall limit our discussion of the probability theory underlying Types A and B to showing in Chapter VII that a certain line of development of the binomial distribution suggests the use of the

Type A series as an extension of the ordinary De Moivre-Laplace approximation, and the Type B series as an extension of the Poisson exponential approximation considered in Chapter II. This development is postponed to the final chapter of the book because it involves more formal mathematics than some readers may find it convenient to follow. Certain important results derived in Chapter VII are stated without proof in §§ 19–21. While a mastery of the details of Chapter VII is not essential to an understanding of the results given in §§ 19–21, the reader who can follow a formal mathematical development without special difficulty may well read Chapter VII at this point instead of reading §§ 19–21. In § 56 of Chapter VII we follow closely the recent work of Wicksell[24] in the development of the forms of the Type A and Type B series. Then in §§ 57–59 we deal with the principles involved in the determination of the parameters in these type forms.

19. **The coefficients of the Type A series expressed in moments of the observed distribution.** If we measure x from the centroid of area as an origin and with units equal to the standard deviation, σ, we may write the Type A series in the form

$$(13)\quad \left\{ \begin{array}{r} F(x)=\phi(x)+a_3\phi^{(3)}(x)+a_4\phi^{(4)}(x)+\dots\dots+a_n\phi^{(n)}(x) \\ +\dots\dots, \end{array} \right.$$

where
$$\phi(x)=\frac{1}{\sigma(2\pi)^{1/2}}e^{-x^2/2},$$

and $\phi^{(n)}(x)$ is the nth derivative of $\phi(x)$ with respect to x.

It will be shown in § 57 that the coefficients a_n for $(n=3, 4, \ldots)$ may then be expressed in the form

$$(14) \qquad a_n = \frac{\sigma(-1)^n}{n!} \int_{-\infty}^{\infty} F(x) H_n(x)\,dx\,,$$

where

$$H_n(x) = x^n - \frac{n(n-1)}{2} x^{n-2} + \frac{n(n-1)(n-2)(n-3)}{2 \cdot 4} x^{n-4} - \cdots$$

is a so-called Hermite polynomial.

To determine a_n numerically, we replace $F(x)$ in (14) by the corresponding observed frequency function $f(x)$, and replace x by x/σ if we measure x with ordinary units (feet, pounds, etc.) instead of using the standard deviation as the unit. Then we may write

$$(15) \qquad a_n = \frac{(-1)^n}{n!} \int_{-\infty}^{\infty} f(x) H_n\left(\frac{x}{\sigma}\right) dx\,.$$

Insert the values of $H_n(x/\sigma)$ for $n=3, 4, 5$ in (15), and we obtain coefficients in terms of moments as follows, using the symbol α_s for the quotient μ_s/σ^s

$$a_3 = -\frac{\mu_3}{\sigma^3 3!} = -\frac{\alpha_3}{3!}$$

$$a_4 = \frac{1}{\sigma^4 4!}(\mu_4 - 3\mu_2^2) = \frac{1}{4!}(\alpha_4 - 3)\,,$$

$$a_5 = -\frac{1}{\sigma^5 5!}(\mu_5 - 10\mu_3\sigma^2) = -\frac{1}{5!}(\alpha_5 - 10\alpha_3)\,.$$

20. **Remarks on two methods of determining the coefficients of the Type A series.** It will be shown in § 57 that formula (14) for any coefficient a_n of the Type A series may be derived by making use of the fact that $\phi^{(n)}(x)$ and the Hermite polynomials $H_n(x)$ form a biorthogonal system. Then as indicated on page 168 we obtain a_n in terms of moments of the observed distribution.

As a second method of obtaining a_n in terms of the moments of the observed distribution $f(x)$, it will be shown in § 58 that the values of the coefficients given in § 19 may be derived by imposing the least-squares criterion[25] that

$$V=\int_{-\infty}^{+\infty}\frac{1}{\phi(x)}[f(x)-F(x)]^2dx \tag{16}$$

shall be a minimum.

21. **The coefficients of the Type B series.** For the Type B series (12), we shall for simplicity limit the determination of coefficients to the first three terms. Moreover, we shall restrict our treatment to a distribution of equally distant ordinates at non-negative integral values of x. Then the problem is to find the coefficients c_0, c_1, c_2 in

$$F(x)=c_0\psi(x)+c_1\Delta\psi(x)+c_2\Delta^2\psi(x) ,$$

where

$$\psi(x)=\frac{e^{-\lambda}\lambda^x}{x!}$$

for $x=0, 1, 2, \ldots$.

By expressing the coefficients in terms of moments of the observed distribution as shown in § 59, we find

$$c_0=1\,, \qquad c_1=0\,, \qquad c_2=\tfrac{1}{2}(\mu_2-\lambda)\,,$$

when λ is taken equal to the arithmetic mean of the given observed values.

22. **Remarks.** With respect to the selection of Type A or Type B of Charlier to represent given numerical data, no criterion corresponding to the Pearson criteria has been given which enables one to distinguish between cases in which to apply one of these types in preference to the other, but Type B applies, in general, to certain decidedly skew distributions; and, in particular, to distributions of variates having a natural lower or upper bound with the modal frequency much nearer to such natural bound than to the other end of the distribution. For example, a frequency distribution of the number dying per month in a city from a minor disease would have the modal value near zero, the natural lower bound.

While the systematic procedure in fitting Charlier curves to data is not so well standardized as the methods used in fitting curves of the Pearson system to data, tables of $\phi(t)$, where t is in units of standard deviation, of its integral from 0 to t, and of its second to eighth derivatives are given to five decimal places for the range $t=0$ to $t=5$ at intervals of .01 by James W. Glover,[8] and tables of the function, its integral and first six derivatives are given by N. R. Jörgensen[26] to seven decimal places for $t=0$ to $t=4$.

23. **Skewness.** Charlier has fittingly called the coefficients $a_3, a_4, a_5, \ldots.$, along with the mean and standard

deviation, the characteristics of the distribution. The coefficients a_3 and a_4 may be interpreted so as to give characteristics which appear very significant in a description of a distribution to a general reader with little or no mathematical training. It is the common experience of those who have dealt with actual distributions of practical statistics that many of the distributions are not symmetrical. A measure is needed to indicate the degree of asymmetry or skewness of distributions in order that we may describe and compare the degrees of skewness of different distributions.

A measure of skewness is given by

$$S=-3a_3=\frac{\mu_3}{2\sigma^3}=\tfrac{1}{2}\alpha_3 . \tag{17}$$

Another measure of skewness is

$$S=\frac{\text{Mean}-\text{Mode}}{\sigma} . \tag{18}$$

In this latter measure we have adopted a convention as to sign by which the skewness is positive when the mean is greater than the mode. Some authors define skewness as equal numerically but opposite in sign to the value in our definition.

We may easily prove that the measures (17) and (18) are equal for a distribution given by the Pearson Type III curve, and approximately equal for a distribution given by the first two terms of the Gram-Charlier Type A when S as defined in (17) is not very large.

For the Pearson Type III (p. 54),

$$\frac{dy}{dx}=\frac{(x+a)y}{c_0+c_1x} .$$

When the parameters in this equation are expressed in moments about the mean, the equation takes the form

$$\frac{1}{y}\frac{dy}{dx}=\frac{x+\mu_3/2\sigma^2}{\mu_2+\mu_3 x/2\sigma^2},$$

if the origin is at the mean of the distribution. The mode is the value of x for which

$$\frac{dy}{dx}=0 \text{ or } x=-\frac{\mu_3}{2\sigma^2}.$$

That is,

$$\frac{\text{Mean}-\text{Mode}}{\sigma}=\frac{\mu_3}{2\sigma^3}.$$

Hence the measures (17) and (18) are equal for the Type III distribution.

For a distribution given by the first two terms of Type A, we are to consider the frequency curve

$$\text{(19)}\quad \begin{cases} y=\phi(x)+a_3\phi^{(3)}(x) \\ =\dfrac{1}{\sigma(2\pi)^{1/2}}\left[1-\dfrac{\mu_3}{2\sigma^3}\left(\dfrac{x}{\sigma}-\dfrac{x^3}{3\sigma^3}\right)\right]e^{-\frac{x^2}{2\sigma^2}} \\ =\dfrac{1}{\sigma(2\pi)^{1/2}}\left[1-S\left(\dfrac{x}{\sigma}-\dfrac{x^3}{3\sigma^3}\right)\right]e^{-\frac{x^2}{2\sigma^2}}. \end{cases}$$

We shall now prove that the distance from the mean (origin) to the mode is approximately $-\sigma S$ when S is fairly small.

We have from (19)

$$\log[y\sigma(2\pi)^{1/2}] = -\frac{x^2}{2\sigma^2} + \log\left[1 - S\left(\frac{x}{\sigma} - \frac{x^3}{3\sigma^3}\right)\right]$$

$$= -\frac{x^2}{2\sigma^2} - S\left(\frac{x}{\sigma} - \frac{x^3}{3\sigma^3}\right),$$

if we neglect terms in S^2. Then

$$\frac{1}{y}\frac{dy}{dx} = -\frac{x}{\sigma^2} - \frac{S}{\sigma} + \frac{x^2 S}{\sigma^3} = 0$$

at the mode. Solving the quadratic for x we obtain $x = -\sigma S$ if we neglect terms of the order S^3. Hence, the measures (17) and (18) are approximately equal for a distribution given by the first two terms of a Gram-Charlier Type A series.

24. **Excess.** In the general description of a given frequency distribution, we may add an important feature to the description by considering the relative number of variates in the immediate neighborhood of some central value such as the mean or the mode. That is, it would add to the description to give a measure of the degree of peakedness of a frequency curve fitted to a distribution by comparison with the corresponding normal curve fitted to the same distribution. The measure of the peakedness to which we shall now give attention is sometimes called the excess and sometimes the measure of kurtosis.

The excess or degree of kurtosis is measured by

$$E = 3a_4 = \tfrac{1}{8}\left(\frac{\mu_4}{\sigma^4} - 3\right) = \tfrac{1}{8}(\alpha_4 - 3).$$

If the excess is positive, the number of variates in the neighborhood of the mean is greater than in a normal distribution. That is, the frequency curve is higher or more peaked in the neighborhood of the mean than the corresponding normal curve with the same standard deviation. On the other hand, if the excess is negative, the curve is more flat topped than the corresponding normal curve. To obtain a clearer insight into the relation of the measure of excess to the theoretical representation of frequency, let us consider a Gram-Charlier series of Type A to three terms

$$(20)\quad \begin{cases} y=\phi(x)+a_3\phi^{(3)}(x)+a_4\phi^{(4)}(x) \\ \displaystyle =\frac{e^{-\frac{x^2}{2\sigma^2}}}{\sigma(2\pi)^{1/2}}\left[1-\frac{\mu_3}{6\sigma^3}\left(\frac{3x}{\sigma}-\frac{x^3}{\sigma^3}\right)\right. \\ \displaystyle \qquad\qquad\left.+\frac{1}{24}\left(\frac{\mu_4}{\sigma^4}-3\right)\left(\frac{x^4}{\sigma^4}-\frac{6x^2}{\sigma^2}+3\right)\right] \\ \displaystyle =\frac{e^{-\frac{x^2}{2\sigma^2}}}{\sigma(2\pi)^{1/2}}\left[1-\frac{\mu_3}{6\sigma^3}\left(\frac{3x}{\sigma}-\frac{x^3}{\sigma^3}\right)+\frac{E}{3}\left(\frac{x^4}{\sigma^4}-\frac{6x^2}{\sigma^2}+3\right)\right] \end{cases}$$

When we compare the ordinate of (20) at the mean $x=0$ with the ordinate $1/\sigma(2\pi)^{1/2}$ at the mean for the normal curve, we observe that this ordinate exceeds the corresponding ordinate of the normal curve by $E/\sigma(2\pi)^{1/2}$. That is, the excess E is equal to the coefficient by which to multiply the ordinate at the centroid of the normal curve to get the increment to this ordinate as calculated by retaining the terms in $\phi^{(3)}(x)$ and $\phi^{(4)}(x)$ of the Type A series.

25. **Remarks on the distribution of certain transformed variates.** Underlying our discussion of frequency functions, there has perhaps been an implication that

the various types of distribution could be accounted for by an appropriate theory of probability. There may, however, be other than chance factors that produce significant effects on the type of the distribution. Such effects may in certain cases be traced to their source by regarding the variates of a distribution as the results of transformations of the variates of some other type of distribution. Edgeworth was prominent in thus regarding certain distributions. For simple examples, we may think of the diameters, surfaces, and volumes of spheres that represent objects in nature, such as oranges on a tree or peas on a plant. Suppose the distribution of diameters is a normal distribution. It seems natural to inquire into the nature of the distribution of the corresponding surfaces and volumes. The partial answer to the inquiry is that these are distributions of positive skewness. The same kind of problem would arise if we knew that velocities, v, of molecules of gas were normally distributed, and were required to investigate the distribution of energies $mv^2/2$.

To illustrate somewhat more concretely with actual data it may be observed in looking over the frequency distributions of the various subgroups on build of men, in Volume I of the *Medico-Actuarial Mortality Investigation*, that the distributions with respect to weight are, in general, not so nearly symmetrical as the distributions as to height. In fact, the distributions as to weight exhibit marked positive skewness. For example, in the age group 25 to 29 and height 5 feet 6 inches we find the following distribution:

W..........	105	120	135	150	165	180	195	210
F..........	17	722	2,175	1,346	485	155	33	3,

Where W = weight in pounds, F = frequency.

A similar feature had been observed by the writer in examining many frequency distributions of ears of corn with respect to length of ears and weight of ears. The distributions as to weight showed this tendency to positive skewness, whereas the distributions as to lengths of ears were much more nearly symmetrical. It seems natural to assume that the weights of bodies are closely correlated with volumes. We may next take account of the fact that volumes of similar solids vary as the cubes of like linear dimensions.

Such concrete illustrations suggest the investigation of the equation of the frequency curve of values obtained by the transformation of variates of a normal distribution by replacing each variate x of the normal distribution by an assigned function of the form kx^n, where k is a positive constant and n is a positive integer or the reciprocal of a positive integer. A paper on this subject by the writer appeared in the *Annals of Mathematics*[27] in June, 1922. The skewness observed in the distributions of weights is similar to the skewness which results as the effect of this transformation when n is a positive constant.

From a different standpoint S. D. Wicksell[28] in the *Arkiv för Matematik, Astronomi, och Fysik in 1917* has discussed, by means of a generalized hypothesis about elementary errors, a connection between certain functions of a variate and a genetic theory of frequency. The hypotheses involved in this theory are at least plausible in their relation to certain statistical phenomena. There are thus at least two points of view which indicate that the method which uses variates resulting from transformation may rise above the position of a device for fitting distributions and be given a place in the theory of frequency. A recent paper[29] by E. L. Dodd presents a somewhat critical study of the determination of the frequency law of a function of variables with given frequency laws, and another recent paper[30] by S. Bernstein deals with appropriate transformations of variates of certain skew distributions.

26. **Remarks on the use of various frequency functions as generating functions in a series representation.** In the *Handbook of Mathematical Statistics* (1924), page 116, H. C. Carver called attention to certain generating functions designed to make frequency series more rapidly convergent than the Type A series. In a paper published in 1924 on the "Generalization of Some Types of the Frequency Curves of Professor Pearson" (*Biometrika*, pp. 106–16), Romanovsky has used Pearson's frequency functions of Types I, II, and III as the generating functions of infinite series in which these types are involved in a manner analogous to the way in which the normal probability function is involved in the Gram-Charlier series.

When Type I,

$$y = y_0\left(1+\frac{x}{a}\right)^{\alpha}\left(1-\frac{x}{b}\right)^{\beta} = \frac{y_0 u_0}{a^{\alpha} b^{\beta}},$$

is used as a generating function, certain functions ϕ_k, which are polynomials of Jacobi in slightly modified form, occur in the expansion in a way analogous to that in which the Hermite polynomials occur in the Gram-Charlier expansion. Moreover, the analogy is continued because $u_0\phi_k$ and ϕ_k form a biorthogonal system, and this property facilitates the determinations of the coefficients in the series.

When the Type III function

$$y=y_0\left(1+\frac{x}{a}\right)^{\nu a}e^{-\nu x}$$

is used as a generating function, certain functions ϕ_k, which are polynomials of Laguerre in generalized form, play a rôle similar to that of the polynomials of Hermite in the Gram-Charlier expansion.

While it is at least of theoretical interest that various frequency functions may assume rôles in the series representation of frequency somewhat similar to the rôle of the normal frequency function in the Gram-Charlier theory, the fact should not be overlooked that the usefulness of any series representation in applications to numerical data is much restricted by the requirement of such rapid convergence of the series that only a few terms need be taken to obtain a useful approximation.

CHAPTER IV

CORRELATION

27. The meaning of simple correlation. Suppose we have data consisting of N pairs of corresponding variates (x_i, y_i), $i=1, 2, \ldots, N$. The given pairs of values may arise from any one of a great variety of situations. For example, we may have a group of men in which x represents the height of a man and y his weight; we may have a group of fathers and their oldest sons in which x is the stature of a father and y that of his oldest son; we may have minimal daily temperatures in which x is the minimal daily temperature at New York and y the corresponding value for Chicago; we may be considering the effect of nitrogen on wheat yield where x is pounds of nitrogen applied per acre and y the wheat yield; we may be throwing two dice where x is the number thrown with the first die and y the number thrown with the two dice together. If such a set of pairs of variates is represented by dots marking the points whose rectangular co-ordinates are (x, y), we obtain a so-called "scatter-diagram."

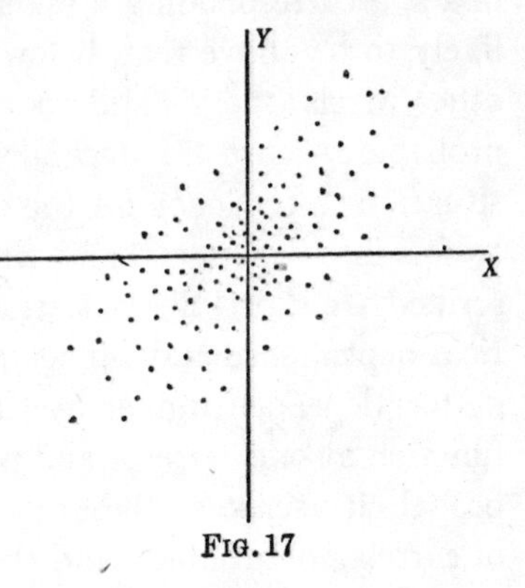

FIG. 17

Assume next that we are interested in a quantitative

characterization of the association of the x's and the corresponding y's. One of the most important questions which can be considered in such a characterization is that of the connection or correlation as it is called between the two sets of values. It is fairly obvious from the scatter-diagram that, with values of x in an assigned interval dx (dx small), the corresponding values of y may differ considerably and thus the y corresponding to an assigned x cannot be given by the use of a single-valued function of x. On the other hand, it may be easily shown that in certain cases, for an assigned x larger than the mean value of x's, a corresponding y taken at random is much more likely to be above than below the mean value of y's. In other words, the x's and y's are not independent in the probability sense of independence. There is often in such situations a tendency for the dots of the scatter-diagram to fall into a sort of band which can be fairly well described. In short, there exists an important field of statistical dependence and connection between the regions of perfect dependence given by a single-valued mathematical function at one extreme and perfect independence in the probability sense at the other extreme. This is the field of correlated variables, and the problems in this field are so varied in their character that the theory of correlation may properly be regarded as an extensive branch of modern methodology.

28. **The regressive method and the correlation surface method of describing correlation.** It may help to visualize the theory of correlation if we point out two fundamental ways of approach to the characterization of a distribution of correlated variables, although the two methods have much in common. The one may be called

the "regression method," and the other the "correlation surface method."

Let us assume that the pairs of variates (x, y) are represented by dots of a scatter-diagram, and set the problem of characterizing the correlation. First, separate the dots into classes by selecting class intervals dx. When we restrict the x's to values in such an interval dx, the set of corresponding y's is called an x-array of y's or simply an array of y's. Similarly, when we restrict the assignment of y's to a class interval dy, the corresponding set of x's is called a y-array of x's or simply an array of x's. The whole set of arrays of a variable, say of y, is often called a set of parallel arrays.

The *regression curve* $y=f(x)$ of y on x for a population is defined to be the locus of the expected value (§ 6) of the variable y in the array which corresponds to an assigned value of x, as dx approaches zero. In other words, the regression curve of y on x is the locus of the means of arrays of y's of the theoretical distribution, as dx approaches zero.

These equivalent definitions relate to the ideal population from which a sample is to be drawn. The regression curve found from a sample is merely a numerical approximation to the ideal set up in the definition.

In the regression method, our first interest is in the regression curves of y on x and of x on y. We are interested next in the characterization of the distribution of the values of y (array of y's) whose expected or average value we have predicted. This is accomplished to some extent by means of measures of dispersion of the values of y which correspond to an assigned value of x. To illustrate the regression method by reference to the correlation between statures of father and son, we may say that the first concern in the use of the regression method is with predicting the mean stature of a subgroup of men whose fathers are of any assigned height, and the next concern is with predicting the dispersion of such a subgroup. The complete characterization of the theoretical distributions underlying arrays of y's may be regarded as the complete solution of the problem of the statistical dependence of y on x.

In the correlation surface method for the two variables, our primary interest is in the characterization of the probability $\phi(x, y)dx\,dy$ that a pair of corresponding variates (x, y) taken at random will fall into the assigned rectangular area bounded by x to $x+dx$ and y to $y+dy$. This method may be regarded as an extension to functions of two or more variables of the method of theoretical frequency functions of one variable. To get at the meaning of correlation by this method, suppose that a function $g(x)$ is such that $g(x)dx$ gives, to within infinitesimals of higher order, the probability that a variate x taken at random lies between x and $x+dx$; and suppose that $h(x, y)dy$ gives similarly the probability that a variate y taken at random from the array of values which correspond to values of x in the interval x to $x+dx$ will lie between y and $y+dy$. Then the probability that the two events will both happen is given by the product

$$\phi(x, y)dx\,dy = g(x)h(x, y)dx\,dy\,. \tag{1}$$

For the probability that both of two events will happen is the product of the probability that the first will happen, multiplied by the probability that the second will happen when the first is known to have happened.

Two cases occur in considering this product. In the first case, $h(x, y)$ is a function of y alone. When this is the case we say the x and y variates are uncorrelated and $\phi(x, y)$ is simply the product of a function of x only multiplied by a function of y only. In such a case the probability that a variate y will be between y and $y+dy$ is the same whether the corresponding assigned x be large or small. In the second case $h(x, y)$ is a function of both x and y. In such cases, the probability that a variate y will be between y and $y+dy$ is not, in general, the same for corresponding assigned large and small values of x. In such cases the two systems of variates are said to be correlated. Thus, in considering for example a group of college students, the height of a student is probably uncorrelated with the grades he makes in mathematics or with the income of his father, but his height is correlated with his weight, and with the height of his father.

Both the regression method and the correlation surface method of dealing with correlation have been in evidence almost from the earliest contributions to the subject. The early method of Francis Galton was essentially the regression method, but the mathematical solution of the special problem[31] which he proposed to J. D. Hamilton Dickson in 1886 consisted in giving the equation of the normal frequency surface to correspond to given lines of regression. The solution of this problem thus involved the correlation surface method. Furthermore, the early contributions of Karl Pearson to correlation theory, involving the influence of selection, stressed frequency surfaces[32] more than regression equations. But, beginning with a paper[33] by G. Udny Yule in 1897, the theory has been developed without limitation to a particular type of

That is, the correlation coefficient of two sets of variates, expressed with their respective standard deviations as units, may be defined as the arithmetic mean of the products of deviations of corresponding values from their respective arithmetic means.

We have defined the correlation coefficient r for a sample. The expected value of the right-hand member of (2) in the sampled population is then the *correlation coefficient for the population.*

While the formula (2) is very useful for the purpose of giving the meaning of the correlation coefficient, other formulas easily obtained from (2) are usually much better adapted to numerical computation. For example,

$$r = \frac{\frac{1}{N}\sum x_i y_i - \bar{x}\bar{y}}{\sigma_x \sigma_y}, \tag{3}$$

$$r = \frac{\frac{1}{N}\sum x_i y_i - \bar{x}\bar{y}}{\left[\frac{1}{N}\sum x_i^2 - \bar{x}^2\right]^{1/2}\left[\frac{1}{N}\sum y_i^2 - \bar{y}^2\right]^{1/2}} \tag{4}$$

are ordinarily more convenient than (2) for purposes of computation.

When N is small, say <30, formula (4) is readily applied. When N is large, appropriate forms for the calculation of r are available in various books.

Still other forms for expressing r are useful for certain purposes. For example, for the purpose of showing that $-1 \leqq r \leqq 1$, we shall now give two further formulas for r.

By simple algebraic verification and remembering that

$1=\sum x_i'^2/N=\sum y_i'^2/N$, it follows that (2) may be written in the forms[34]

$$r=1-\frac{1}{2N}\sum(x_i'-y_i')^2\,, \tag{5}$$

$$r=-1+\frac{1}{2N}\sum(x_i'+y_i')^2\,. \tag{6}$$

From these two formulas, we have the important proposition that

$$-1\leqq r\leqq 1\,. \tag{7}$$

THE REGRESSION METHOD OF DESCRIPTION

30. **Linear regression.** Suppose we are interested in the mean value $\bar{y}_x$ of the y's in the x-array of y's. The simplest and most important case to consider from the standpoint of the practical problems of statistics is that in which the regression of y on x is a straight line. Assuming that the regression curve of y on x in the population is a straight line, we accept as an approximation the line $y_x=mx+b$ which fits "best" the means of arrays of the sample.

The term "best" is here used to mean best under a least-squares criterion of approximation. In applying the criterion the square $(\bar{y}_x-mx-b)^2$ for each array is weighted with the number in the array. Let N_x be the number of dots in any assigned x-array of y's. Then the equation of our line of regression would be

$$\bar{y}_x=mx+b\,, \tag{8}$$

where m and b are to be determined by the condition that the sum

$$\sum N_x(\bar{y}_x-mx-b)^2\,, \tag{9}$$

with observed data substituted for x, $\bar{y}_x$ and N_x from all arrays, is to be a minimum. Differentiating (9) with respect to b and m, we have

$$(10) \qquad -2\sum N_x(\bar{y}_x - mx - b) = 0\,,$$

$$(11) \qquad -2\sum N_x(\bar{y}_x - mx - b)x = 0\,.$$

We may note that $N_x\bar{y}_x$ is equal to the sum of all y's in an array of y's. If we examine these equations on making substitutions for $\bar{y}_x$ and x, it is easily seen that they are, except for grouping errors which vanish as $dx \to 0$, equivalent to the equations

$$(12) \qquad -2\sum (y_i - mx_i - b) = 0\,,$$

$$(13) \qquad -2\sum x_i(y_i - mx_i - b) = 0\,,$$

where the summation is extended to all the given pairs. That is, we may find the regression line by obtaining the linear function $y = mx + b$, which gives the best least-square estimate of the values of y which correspond to assigned values of x. Take the origin at the mean of x's and the mean of y's. Then $\sum y_i = 0$, $\sum x_i = 0$. Hence, from (12), $b = 0$. From (13)

$$m = \frac{\sum x_i y_i}{\sum x_i^2} = \frac{\sum x_i y_i}{N\sigma_x\sigma_y} \cdot \frac{\sigma_y}{\sigma_x} = r\,\frac{\sigma_y}{\sigma_x}\,,$$

and the equation of the line of regression of y on x is

$$(14) \qquad y = r\,\frac{\sigma_y}{\sigma_x}\,x\,.$$

Similarly, the line of regression of x on y is

$$(15) \qquad x = r \frac{\sigma_x}{\sigma_y} y .$$

It should be remembered that the origin is at the mean values of x's and of y's when the regression equations take the forms (14) and (15). It is obvious that these equations may be written as

$$(16) \qquad y - \bar{y} = r \frac{\sigma_y}{\sigma_x} (x - \bar{x})$$

and

$$(17) \qquad x - \bar{x} = r \frac{\sigma_x}{\sigma_y} (y - \bar{y})$$

when we take any arbitrary origin.

The coefficient $r\sigma_y/\sigma_x$ is called the regression coefficient of y on x, and similarly $r\sigma_x/\sigma_y$ is the regression coefficient of x on y.

If we use standard deviations as units of measurement the regression equations (14) and (15) become

$$(18) \qquad y' = rx' , \quad x' = ry' ,$$

and the regression coefficients are equal to each other and to the correlation coefficient.

When there is no correlation between x's and y's, $r=0$, and the regression lines of y on x and of x on y are parallel to the x- and y-axes, respectively. On the other hand, when $r=0$, it is not necessarily true that there is no correlation. Indeed, there may be a high correlation[35] with non-linear regression when $r=0$. For example, we may have $r=0$ when y is a simple periodic function of x.

31. The standard deviation of arrays—mean square error of estimate. In passing judgment on the degree of precision to be expected in estimating the value of a variable, say y, by means of the regression equation of y on x, it is important to have a measure of the dispersion in arrays of y's.

The mean square error s_y^2 involved in taking the ordinates of the line of regression as the estimated values of y may be very simply expressed by $s_y^2=\sigma_y^2\,(1-r^2)$. To prove that s_y^2 takes this value, we may write the sum of the squares of deviations in the form

$$Ns_y^2=\sum\left(y-r\,\frac{\sigma_y}{\sigma_x}\,x\right)^2=\sum y^2-2r\,\frac{\sigma_y}{\sigma_x}\sum xy+r^2\,\frac{\sigma_y^2}{\sigma_x^2}\sum x^2$$

$$=N\sigma_y^2-2Nr^2\sigma_y^2+Nr^2\sigma_y^2=N\sigma_y^2(1-r^2)\ .$$

Hence, we have

$$s_y^2=\sigma_y^2(1-r^2)\ , \tag{19}$$

$$s_y=\sigma_y(1-r^2)^{1/2}\ . \tag{20}$$

This value of s_y may be regarded as a sort of average value of the standard deviations of the arrays of y's, and is sometimes called the root-mean-square error of estimate of y, or more briefly, *the standard error of estimate* of y. The factor $(1-r^2)^{1/2}$ in (20) has been called the *coefficient of alienation* or the measure of the failure to improve the estimate of y from knowledge of the correlation.

When the standard deviation of an array of y's is regarded as a function, say $S(x)$, of the assigned x, the curve $y=S(x)/\sigma_y$ is called the *scedastic curve.* It may be described as the curve whose ordinates measure the scatter in arrays of y's in comparison to the scatter of all y's. When $S(x)$ is a constant, the regression system of y on x is called a *homoscedastic system.* When $S(x)$ is not constant, the system is said to be *heteroscedastic.* For a homoscedastic system with linear regression, $s_y = \sigma_y(1-r^2)^{1/2}$ is the standard deviation of each array of y's.

To illustrate (20) numerically, let us suppose that $r = .5$ gives the correlation of statures of fathers and sons. Assuming linear regression, the root-mean-square error of estimate of the height of a son derived from the assigned height of the father would be

$$s_y = \sigma_y[.75]^{1/2} = .866\sigma_y .$$

That is, the average dispersion in the arrays of heights of sons which correspond to assigned heights of fathers is about .87 as great as the dispersion of the heights of all the sons. It is, therefore, fairly obvious that we cannot, with any considerable degree of reliability, predict from $r = .5$ the height of an individual son from the height of the father. However, with a large N, we can give a very reliable prediction of the mean heights of sons that correspond to assigned heights of fathers.

It should be remembered that we have thus far assumed linear regression of y on x. An analogous consideration of the dispersion in arrays of x's gives for the mean square error of estimate

$$s_x^2 = \sigma_x^2(1-r^2)$$

when we assume linear regression of x on y.

32. Non-linear regression—the correlation ratio. In case a curve of regression, say of y on x, is not a straight line, the correlation coefficient as a measure of correlation may be misleading. In introducing a correlation ratio, η_{yx}, of y on x as an appropriate measure of correlation to take the place of the correlation coefficient in such a situation, we may get suggestions as to what is appropriate by solving for r^2 in (19). This gives

$$r^2 = 1 - s_y^2/\sigma_y^2 , \tag{21}$$

where we may recall that s_y^2 is the mean square of deviations from the line of regression. Then

$$r = \pm(1 - s_y^2/\sigma_y^2)^{1/2} .$$

This formula could be used appropriately as a definition of r in place of our definition in (2), and its examination may throw further light on the significance of r. When $s_y = 0$, the formula gives $r = 1$ and, as we have seen earlier, all the dots of the scatter-diagram must then fall exactly on the line of regression $y = r\sigma_y x/\sigma_x$. When $s_y = \sigma_y$, the formula gives $r = 0$, and the regression line is in this case of no aid in predicting the value of y from assigned values of x. In the formula $r^2 = 1 - s_y^2/\sigma_y^2$ it is important to keep in mind that the mean square deviation s_y^2 is from the line of regression (§ 31). Next, let $s_y'^2$ be the corresponding mean square of deviations from the means of arrays. Then in the population $s_y'^2 = s_y^2$ when the regression is strictly linear, but $s_y'^2 \neq s_y^2$ when the regression is non-linear. This fact suggests the use of a formula closely related to $[1 - s_y^2/\sigma_y^2]^{1/2}$ for a measure of non-linear regression by replacing s_y by s_y'. We then write

$$\eta_{yx}^2 = 1 - s_y'^2/\sigma_y^2 , \tag{22}$$

where η_{yx} is the *correlation ratio* of y on x, and $s_y'^2$ is the mean square of deviations from the means of arrays whether these means are near to or far from the line of regression. For linear regression of y on x, we have $\eta_{yx}^2 = r^2$ in the population.

In general, we may say that the correlation ratio of y on x is a measure of the clustering of dots of the scatter-diagram about the means of arrays of y's.

An analogous discussion for the arrays of x's obviously leads to

$$\eta_{xy}^2 = 1 - s_x'^2/\sigma_x^2 ,$$

giving η_{xy}, the correlation ratio of x on y.

That $\eta_{yx}^2 \leqq 1$ and that the equality holds only when all the dots in each array are at the mean of the array follows at once from (22).

That $\eta_{yx}^2 \geqq r^2$ may be shown by recalling the meanings of s_y^2 in (21) and of $s_y'^2$ in (22). A mean square of deviations in each array is a minimum when the deviations are taken from the mean of the array. Hence, the $s_y'^2$ in (22) must be equal to or less than s_y^2 in (21) for the same data, since the deviations in (21) are measured from the line of regression. Hence, we have shown that

$$1 \geqq \eta_{yx}^2 \geqq r^2 .$$

Moreover, when the regression of y on x is linear, $\eta_{yx}^2 - r^2$ found from the sample differs from zero by an amount not greater than the fluctuations due to random sampling. Indeed, the comparison of the quantity $\eta_{yx}^2 - r^2$ with its sampling errors becomes the most useful known criterion for testing the linearity of the regression of y on x.

For some purposes, it is convenient to express the correlation ratios in a form involving the standard deviation of the means of arrays. For this purpose, let $\bar{y}_x$ be the mean of any array of y's, and $\sigma_{\bar{y}_x}$ the standard deviation of the means of arrays when the square $(\bar{y}_x - \bar{y})^2$ of each deviation is weighted with the number N_x in the array. Then it follows very simply that

$$\eta_{yx}^2 = \frac{\sigma_y^2 - s_y'^2}{\sigma_y^2} = \frac{\sigma_{\bar{y}_x}^2}{\sigma_y^2} . \tag{23}$$

That is, the correlation ratio of y on x is the ratio of the standard deviation of the means of arrays of y's to the standard deviation of all y's.

The calculation of the correlation ratio with a large number N of pairs may be carried out very conveniently as a mere extension of the calculation of the correlation coefficient. For a form for such calculation, see *Handbook of Mathematical Statistics*, page 130.

In order to get a fair approximation to a correlation ratio in a population from a sample, it is important that the grouping into class intervals be not so narrow as to give arrays containing very few variates. Certain valuable formulas for the correction of errors due to grouping have been published.[36]

When the regression is non-linear, the correlation may be further characterized by the equation of a curve of regression that passes approximately through the means of arrays of a given system of variates. As early as 1905, the parameters of the special regression curves given by polynomials $y = f(x)$ of the second and third degrees were determined in terms of power moments and product moments. In 1921, Karl Pearson[37] published a general method of determining successive terms of the regression curve of the form

$$y = f(x) = a_0\psi_0 + a_1\psi_1 + \cdots + a_n\psi_n, \qquad (24)$$

where $a_0, a_1, \ldots, a_n$ are constants to be determined and the functions ψ_s form an orthogonal system of functions of x. That is,

$$\sum (N_x \psi_s \psi_{s'}) = 0,$$

if the summation $\sum$ be taken for all values of x corresponding to a system of arrays with frequency in an x-array given by N_x. An exposition of the theory of non-linear regression curves is somewhat beyond the scope of this monograph.

33. **Multiple correlation.** Thus far we have considered only simple correlation, that is, correlation between two variables. But situations frequently arise which call for the investigation of correlation among three or more variables. A familiar example occurs in the correlation of a character such as stature in man with statures of each of the two parents, of each of the four grandparents, and possibly with statures of others back in the ancestral line. Other examples can be readily cited. Indeed, it is very generally true that several variables enter into many problems of biology, economics, psychology, and education.

The solution of these problems calls for a development of correlation among three or more variables. Suppose we have given N sets of corresponding values of n variables $x_1, x_2, \ldots, x_n$. Assume next that we separate the values of x_1 into classes by selecting class intervals dx_2, dx_3, ... , dx_n of the remaining variables. When we limit the x_2's to an assigned interval dx_2, x_3's to an assigned interval dx_3, and so on, the set of corresponding x_1's is sometimes called an array of x_1's.

The locus of the means of such arrays of x_1's in the theoretical distribution, as $dx_2, dx_3, \ldots , dx_n$ approach zero, is called the *regression surface* of x_1 on the remaining variables. It will be convenient to assume that any variable, x_j, is measured from the arithmetic mean of its N given values as an origin. Let σ_j be the standard deviation of the N values of x_j, and let r_{pq} be the correlation coefficient of the N given pairs of values of x_p and x_q. Then we seek to determine $b_{12}, b_{13}, \ldots , b_{1n}, c$, the parameters in the linear regression surface,

$$x_1 = b_{12}x_2 + b_{13}x_3 + \cdots + b_{1n}x_n + c \,, \tag{25}$$

of x_1 on the remaining variables so that x_1 computed from (25) will give on the whole the "best" estimates of the values of x_1 that correspond to any assigned values of $x_2, x_3, \ldots , x_n$. Adopting a least-squares criterion, we may determine the coefficients in (25) so that

$$U = \sum (x_1 - b_{12}x_2 - b_{13}x_3 - \cdots - b_{1n}x_n - c)^2 \tag{26}$$

shall be a minimum. This gives for the linear regression surface of x_1 on $x_2, x_3, \ldots , x_n$,

$$x_1 = -\sigma_1 \sum_{q=2}^{q=n} \frac{R_{1q}}{R_{11}} \frac{x_q}{\sigma_q} \,, \tag{27}$$

where R_{pq} is the cofactor of the pth row and the qth column of the determinant

$$(28)\qquad R=\begin{vmatrix} 1 & r_{12} & r_{13} & \dots & r_{1n} \\ r_{21} & 1 & r_{23} & \dots & r_{2n} \\ r_{31} & r_{32} & 1 & \dots & \\ \vdots & & & & \\ r_{n1} & r_{n2} & & \dots & 1 \end{vmatrix}$$

For simplicity we shall limit ourselves to $n=3$ in giving proofs of these statements, but the method can be extended in a fairly obvious manner from three variables to any number of variables.

Equating to zero the first derivatives of U in (26) with respect to c, b_{12}, and b_{13}, we obtain, when $n=3$, the equations

$$c=0\,,$$

$$-2\sum x_2(x_1-b_{12}x_2-b_{13}x_3)=0\,.$$

$$-2\sum x_3(x_1-b_{12}x_2-b_{13}x_3)=0\,,$$

The last two equations may be written in the form

$$\sum x_1x_2-b_{12}\sum x_2^2-b_{13}\sum x_2x_3=0\,,$$

$$\sum x_1x_3-b_{12}\sum x_2x_3-b_{13}\sum x_3^2=0\,.$$

By expressing the summations in terms of standard deviations and correlation coefficients, we have

$$(29)\qquad Nb_{12}\sigma_2^2+Nb_{13}r_{23}\sigma_2\sigma_3=Nr_{12}\sigma_1\sigma_2\,,$$

$$(30)\qquad Nb_{12}r_{23}\sigma_2\sigma_3+Nb_{13}\sigma_3^2=Nr_{13}\sigma_1\sigma_3\,.$$

Solving for b_{12} and b_{13}, we obtain

$$b_{12}=\frac{\sigma_1\sigma_2\sigma_3^2}{\sigma_2^2\sigma_3^2}\left|\frac{\begin{matrix} r_{12} & r_{23} \\ r_{13} & 1 \end{matrix}}{\begin{matrix} 1 & r_{23} \\ r_{23} & 1 \end{matrix}}\right|=\frac{\sigma_1}{\sigma_2}\left|\frac{\begin{matrix} r_{12} & r_{23} \\ r_{13} & 1 \end{matrix}}{\begin{matrix} 1 & r_{23} \\ r_{23} & 1 \end{matrix}}\right|,$$

$$b_{13}=\frac{\sigma_1}{\sigma_3}\left|\frac{\begin{matrix} 1 & r_{12} \\ r_{23} & r_{13} \end{matrix}}{\begin{matrix} 1 & r_{23} \\ r_{23} & 1 \end{matrix}}\right|$$

Hence

$$x_1=-\sigma_1\sum_{q=2}^{q=3}\frac{R_{1q}}{R_{11}}\frac{x_q}{\sigma_q},$$

where R_{pq} is the cofactor of the pth row and qth column of

$$R=\begin{vmatrix} 1 & r_{12} & r_{13} \\ r_{21} & 1 & r_{23} \\ r_{31} & r_{32} & 1 \end{vmatrix}$$

If the dispersion (scatter) $\sigma_{1.23\,\ldots\,n}$ of the observed values of x_1 from its corresponding computed values on the hyperplane (27) is defined as the square root of mean square of the deviations, that is,

$$\text{(31)}\qquad \sigma^2_{1.23\,\ldots\,n}=\frac{1}{N}\sum(\text{observed } x_1-\text{computed } x_1)^2,$$

then it can be proved that

$$\text{(32)}\qquad \sigma_{1.23\,\ldots\,n}=\sigma_1(R/R_{11})^{1/2}.$$

To prove this for $n=3$, we may write from (27) and (31)

$$N\sigma_{1.23}^2=\frac{\sigma_1^2}{R_{11}^2}\sum\left(R_{11}\frac{x_1}{\sigma_1}+R_{12}\frac{x_2}{\sigma_2}+R_{13}\frac{x_3}{\sigma_3}\right)^2$$

$$=\frac{\sigma_1^2}{R_{11}^2}N(R_{11}^2+R_{12}^2+R_{13}^2+2R_{11}R_{12}r_{12}+2R_{11}R_{13}r_{13}+2R_{12}R_{13}r_{23})$$

$$=\frac{\sigma_1^2 N}{R_{11}^2}[R_{11}(R_{11}+r_{12}R_{12}+r_{13}R_{13})+R_{12}(R_{12}+r_{12}R_{11}+r_{23}R_{13})$$
$$+R_{13}(R_{13}+r_{13}R_{11}+r_{23}R_{12})]\,.$$

Since from elementary theorems of determinants,

$$R_{11}+r_{12}R_{12}+r_{13}R_{13}=R\,,$$

$$R_{12}+r_{12}R_{11}+r_{23}R_{13}=0\,,$$

$$R_{13}+r_{13}R_{11}+r_{23}R_{12}=0\,,$$

we have

$$(33)\qquad \sigma_{1.23}^2=\sigma_1^2\,R/R_{11}\,,\qquad \sigma_{1.23}=\sigma_1(R/R_{11})^{1/2}\,.$$

As an extension of the standard error of estimate with two variables (p. 87), it is true for n variables that the standard error $\sigma_{1.23\,\ldots\,n}$ of estimating x_1 from assigned values of $x_2, x_3, \ldots, x_n$ is the standard deviation of each array of x_1's, provided all regressions are linear and the standard deviation of an array of x_1's is the same for all sets of assignments of $x_2, x_3, \ldots, x_n$.

Next, we shall inquire into the dispersion of the estimated values given by (27). Since the mean value of these estimates is zero, when the origin is at the mean of each system of variates, we have the standard deviation σ_{1E} of the estimates of x_1 given by

$$\sigma_{1E}^2 = \frac{\sigma_1^2}{N} \sum \left(\frac{R_{12}}{R_{11}} \frac{x_2}{\sigma_2} + \frac{R_{13}}{R_{11}} \frac{x_3}{\sigma_3} \right)^2 = \frac{\sigma_1^2}{R_{11}^2} (R_{12}^2 + R_{13}^2 + 2R_{12}R_{13}r_{23})$$

$$= \frac{-\sigma_1^2}{R_{11}} \{ r_{12}R_{12} + r_{13}R_{13} \} = \sigma_1^2 \left(1 - \frac{R}{R_{11}} \right).$$

The correlation coefficient $r_{1.23 \ldots n}$ between the observed values of x_1 and its corresponding estimated values calculated from the linear function (27) of $x_2, x_3, \ldots, x_n$ is called the *multiple correlation coefficient* of order $n-1$ of x_1 with the other $n-1$ variables. The multiple correlation coefficient $r_{1.23 \ldots n}$ is expressible in terms of simple correlation coefficients by the formula

$$r_{1.23 \ldots n} = [1 - R/R_{11}]^{1/2}. \tag{34}$$

To prove (34), limiting ourselves to $n=3$, we write

$$N\sigma_1\sigma_{1E}r_{1.23} = \sigma_1^2 \sum \frac{x_1}{\sigma_1} \left(-\frac{R_{12}}{R_{11}} \frac{x_2}{\sigma_2} - \frac{R_{13}}{R_{11}} \frac{x_3}{\sigma_3} \right)$$

$$= \frac{-N\sigma_1^2}{R_{11}} (R_{12}r_{12} + R_{13}r_{13})$$

$$= \frac{-N\sigma_1^2}{R_{11}} (R - R_{11}) = N\sigma_1^2(1 - R/R_{11})$$

Since

$$\sigma_{1E} = \sigma_1[1 - R/R_{11}]^{1/2},$$

we have the result sought,

$$r_{1.23} = [1 - R/R_{11}]^{1/2},$$

The relation (34) is very significant because it enables us to express multiple correlation coefficients in terms of simple correlation coefficients.

From equations (32) and (34), it follows that

$$(35) \qquad \sigma^2_{1.23\ldots n} = \sigma^2_1(1 - r^2_{1.23\ldots n}) .$$

34. **Partial correlation.** It is often important to obtain the degree of correlation between two variables x_1 and x_2 when the other variables $x_3, x_4, \ldots, x_n$ have assigned values. For example, we might find the correlation of statures of fathers and sons when the stature of the mother is an assigned constant, say 62 inches. In general, suppose we have found a correlation between characters A and B, and that it is a plausible interpretation that the correlation thus found is due to the correlation of each of them with a character C. In this case we could remove the influence of C, if we had a sufficient amount of data, by restricting our data to a universe of A and B corresponding to an assigned C.

In accord with this notion, we may define a *partial correlation coefficient* $r'_{12.34\ldots n}$ of x_1 and x_2 for assigned $x_3, x_4, \ldots, x_n$, as the correlation coefficient of x_1 and x_2 in the part of the population for which $x_3, x_4, \ldots, x_n$ have assigned values. A change in the selection of assigned values may lead to the same or to different values of $r'_{12.34\ldots n}$.

Suppose we are dealing with a population for which the regression curves are straight lines and the regression surfaces are planes. Thus, let us assume that the theoretical mean or expected values of x_1 and x_2 for an assigned $x_3, x_4, \ldots, x_n$ are

$$b_{13}x_3 + b_{14}x_4 + \cdots + b_{1n}x_n ,$$

$$b_{23}x_3 + b_{24}x_4 + \cdots + b_{2n}x_n ,$$

respectively. Then a partial correlation coefficient $r'_{12.34\ldots n}$ is the simple correlation coefficient of residuals

$$x_{1.34\ldots n}=x_1-b_{13}x_3-b_{14}x_4-\cdots-b_{1n}x_n$$

and

$$x_{2.34\ldots n}=x_2-b_{23}x_3-b_{24}x_4-\cdots-b_{2n}x_n$$

limited to the part of the population $N_{34\ldots n}$ of the total N for which $x_3, x_4, \ldots, x_n$ are fixed.

Suppose further that the population is such that any change in the assignment of values to $x_3, x_4, \ldots, x_n$ does not change the standard deviation of $x_{1.34\ldots n}$. nor of $x_{2.34\ldots n}$, nor the value of $r'_{12.34\ldots n}$. Such a population suggests that we define

$$(36)\qquad r_{12.34\ldots n}=\frac{\sum x_{1.34\ldots n}\,x_{2.34\ldots n}}{N\sigma_{1.34\ldots n}\,\sigma_{2.34\ldots n}},$$

where the summation is extended to N pairs of residuals, as *the partial correlation coefficient* of x_1 and x_2 for all sets of assignments of x_3 $x_4, \ldots, x_n$.

If the population is such that $r'_{12.34\ldots n}$ is not the same for each different set of assignments of $x_3, x_4, \ldots, x_n$, the right-hand member of (36) may still be regarded as a sort of average value of the correlation coefficients of x_1 and x_2 in subdivisions of a population obtained by assigning $x_3, x_4, \ldots, x_n$, or it may be regarded as the correlation coefficient between the deviations of x_1 and x_2 from the corresponding predicted values given by their linear regression equations on $x_3, x_4, \ldots x_n$.

The partial correlation coefficient as given in (36) is

expressible in terms of simple correlation coefficients by the formula

$$r_{12.34\ldots n} = \frac{-R_{12}}{[R_{11}R_{22}]^{1/2}}, \tag{37}$$

where R_{pq} is a cofactor defined in §33.

We may prove (37), limiting ourselves to $n=3$, as follows: By definition

$$r_{12.3} = \frac{\sum x_{1.3}\, x_{2.3}}{N\sigma_{1.3}\,\sigma_{2.3}} = \frac{\sum \left(x_1 - r_{13}\frac{\sigma_1}{\sigma_3}x_3\right)\left(x_2 - r_{23}\frac{\sigma_2}{\sigma_3}x_3\right)}{N\sigma_{1.3}\,\sigma_{2.3}}$$

$$= \frac{\sum x_1x_2 - r_{13}\frac{\sigma_1}{\sigma_3}\sum x_2x_3 - r_{23}\frac{\sigma_2}{\sigma_3}\sum x_1x_3 + r_{13}r_{23}\frac{\sigma_1\sigma_2}{\sigma_3^2}\sum x_3^2}{\left[\sum\left(x_1 - r_{13}\frac{\sigma_1}{\sigma_3}x_3\right)^2 \sum\left(x_2 - r_{23}\frac{\sigma_2}{\sigma_3}x_3\right)^2\right]^{1/2}}$$

$$= \frac{r_{12} - r_{13}r_{23}}{[(1-r_{13}^2)(1-r_{23}^2)]^{1/2}} = \frac{-R_{12}}{[R_{11}R_{22}]^{1/2}}.$$

Thus, (37) is proved for $n=3$.

An important relation between partial and multiple correlation coefficients may now be derived. From (37) we have

$$1 - r_{12.34\ldots n}^2 = \frac{R_{11}R_{22} - R_{12}^2}{R_{11}R_{22}}.$$

By a well-known theorem of determinants,[38]

$$\begin{vmatrix} R_{11} & R_{12} \\ R_{12} & R_{22} \end{vmatrix} = R_{11}R_{22} - R_{12}^2 = RR_{11\,22}.$$

Hence we have

$$1-r_{12.34\ldots n}^2=\frac{RR_{11\,22}}{R_{11}R_{22}}=\frac{\dfrac{R}{R_{11}}}{\dfrac{R_{22}}{R_{11\,22}}}=\frac{1-r_{1.23\ldots n}^2}{1-r_{1.34\ldots n}^2},$$

since from (32) and (35),

$$\frac{R}{R_{11}}=1-r_{1.23\ldots n}^2,$$

and similarly

$$\frac{R_{22}}{R_{11\,22}}=1-r_{1.34\ldots n}^2.$$

Thus we can express the partial correlation coefficient $r_{12.34\ldots n}$ of order $n-2$ (the number of variables held constant) in terms of the multiple correlation coefficient $r_{1.23\ldots n}$ of order $n-1$ and the multiple correlation coefficient $r_{1.34\ldots n}$ of order $n-2$.

35. Non-linear regression in n variables—multiple correlation ratio. The theory of correlation for non-linear regression lends itself to extension to the case of more than two variables as has been demonstrated by the contributions of L. Isserlis[39] and Karl Pearson.[40]

Consider the variables $x_1, x_2, \ldots, x_n$, and fix attention on an array of x_1's which corresponds to assigned values of $x_2, x_3, \ldots, x_n$. Next, let $\bar{x}_{1.23\ldots n}$ be the mean of the values in the array of x_1's and let $\bar{\sigma}_{1.23\ldots n}$ be the standard deviation of these means of arrays of x_1's, where the square of each deviation $\bar{x}_{1.23\ldots n}$ from the mean of x_1's is weighted with the number in the array in finding this standard deviation. Then the multiple correlation ratio $\eta_{1.23\ldots n}$ of x_1 on $x_2, x_3, \ldots, x_n$ may be defined by

$$(38) \qquad \eta^2_{1.23\ldots n} = \frac{\bar{\sigma}^2_{1.23\ldots n}}{\sigma^2_1}.$$

The analogy with the case of the correlation ratio for two variables seems fairly obvious. While the method of computing the multiple correlation ratio $\eta_{1\cdot 23\ldots n}$ is simple in principle, it is unfortunately laborious from the arithmetic standpoint.

36. **Remarks on the place of probability in the regression method.** Thus far we have discussed simple correlation by the regression method without using probabilities in explicit form. To be sure, probability theory is involved in the background. It seems fairly obvious that it would be of fundamental interest to construct urn schemata which would give a meaning to the correlation and regression coefficients in pure chance. In a paper[41] published by the author in 1920, certain urn schemata were devised which give linear regression and very simple values for the correlation coefficient. Other schemata apparently equally simple give non-linear regression. The general plan of the schemata consists in requiring certain elements to be common in successive random drawings. It appears that the construction of such urn schemata will tend to give correlation a place in the elementary theory of probability.

In a recent book[42] by the Russian mathematician, A. A. Tschuprow, an important step has been taken toward connecting the regression method of dealing with correlation more closely with the theory of probability. This is accomplished by a consideration of the underlying definitions and concepts for a priori distributions.

It may be noted that we have not based our development of the regression method on a precise definition of correlation. Instead we have attempted a sort of genetic development. It may at this point be helpful in forming a proper notion of the scope and limitations of the regression method to give a definition of correlation from the regression viewpoint. It seems that a general definition will involve probabilities because we shall almost surely wish to idealize actual distributions into theoretical distributions or laws of frequency for purposes of definition. In a general sense, we may say that y is correlated with x whenever the theoretical distributions in arrays of y's are not identical for all possible assigned values of x, and we say that y is uncorrelated with x whenever the theoretical distributions in arrays of y's are identical with each other for all possible values of x. By the identity of the theoretical distributions in arrays of y's, we mean that they have equal means, standard deviations, and other parameters required to characterize completely the distributions. It is fairly obvious that our discussion of the regression method is incomplete in a sense because we have not given a complete characterization of distributions in arrays. Our characterization of the statistical dependence of y on x may be regarded as complete when the arrays of y's are normal distributions, because the distributions are then completely characterized by their arithmetic means and standard deviations.

THE CORRELATION SURFACE METHOD OF DESCRIPTION

37. The normal correlation surfaces. The function

$$z=f(x_1, x_2, \ldots, x_n)$$

is called a *frequency function* of the n variables, $x_1, x_2, \ldots, x_n$, if

$$z\,dx_1\,dx_2 \ldots dx_n$$

gives, to within infinitesimals of higher order, the probability that a set of values of $x_1, x_2, \ldots, x_n$ taken at random will lie in the infinitesimal region bounded by x_1 and x_1+dx_1, x_2 and $x_2+dx_2, \ldots, x_n$ and x_n+dx_n. When the variables are not independent in the probability sense, the surface represented by $z=f(x_1, x_2, \ldots, x_n)$ is called a *correlation surface.*

With the notation of § 29 for simple correlation, the natural extension of the theory underlying the normal frequency function of one variable to functions of two variables x and y leads to the correlation surface

$$z=\frac{1}{2\pi\sigma_x\sigma_y(1-r^2)^{1/2}}\, e^{-\frac{1}{2(1-r^2)}\left(\frac{x^2}{\sigma_x^2}+\frac{y^2}{\sigma_y^2}-\frac{2rxy}{\sigma_x\sigma_y}\right)}.$$

Moreover, with the notation of § 33 on multiple correlation the natural extension to the case of a function of n normally correlated variables $x_1, x_2, \ldots, x_n$ gives a frequency function of the exponential type

$$z=z_0\, e^{-\frac{1}{2}\phi},$$

where ϕ is a homogeneous quadratic function of the n variables which may be written in the form

$$\phi=\frac{1}{R}\left(R_{11}\,\frac{x_1^2}{\sigma_1^2}+R_{22}\,\frac{x_2^2}{\sigma_2^2}+\cdots+2R_{12}\,\frac{x_1}{\sigma_1}\,\frac{x_2}{\sigma_2}+\cdots\right),$$

the determinant R and its cofactors R_{pp} and R_{pq} being defined in § 33. We thus have a correlation surface in space of $n+1$ dimensions.

For purposes of simplicity we shall limit our derivations of normal frequency functions to functions of two and three variables thus restricting the geometry involved to space of three and four dimensions.

The equation of the normal frequency surface may be derived from various sets of assumptions analogous to and extensions of sets of assumptions from which the normal frequency curve may be derived. Some of these derivations make no explicit use of the fact that in normal correlation the regression is linear. That is, linear regression is considered as a property of the frequency surface obtained from other assumptions. But we may connect the frequency-surface method closely with the regression method by involving linear regression of one of the variables on the others as one of the assumptions from which to derive the surface. This is the plan we shall adopt in the following derivation. Let us assume, first, that one set of variates, say the x's, are distributed normally about their mean value taken as an origin. Then in our notation (p. 47 and § 29)

$$\frac{1}{\sigma_x(2\pi)^{1/2}} e^{-\frac{x^2}{2\sigma_x^2}} dx , \tag{39}$$

to within infinitesimals of higher order, is the probability that an x taken at random will lie in the interval dx. Assume next that any array of y's corresponding to an assigned x is a normal distribution with the standard deviation of an array given by $\sigma_y(1-r^2)^{1/2}$ as found earlier in this chapter (§ 31), and finally, assume that the regression of y on x is linear. Then in the notation of simple correlation

$$(40) \qquad \frac{1}{\sigma_y[2\pi(1-r^2)]^{1/2}} e^{-\frac{1}{2\sigma_y^2(1-r^2)}\left(y-r\frac{\sigma_y}{\sigma_x}x\right)^2} dy$$

is, to within infinitesimals of higher order, the probability that a y taken at random from an assigned array of y's will lie in the interval dy.

By using the elementary principle that the probability that both of two events will occur is equal to the product of the probabilities that the first will occur and that the second will occur when the first is known to have occurred, we have the product $z\,dxdy$ of (39) and (40) for the probability, to within infinitesimals of higher order, that x will fall in dx and the corresponding y in dy, where

$$(41) \qquad z=\frac{1}{2\pi\sigma_x\sigma_y(1-r^2)^{1/2}} e^{-\frac{1}{2(1-r^2)}\left(\frac{x^2}{\sigma_x^2}+\frac{y^2}{\sigma_y^2}-\frac{2rxy}{\sigma_x\sigma_y}\right)}$$

is the normal correlation surface in three dimensions.

Let us turn next to the derivation of the normal correlation surface in four dimensions. Following the notation of multiple correlation we seek a normal frequency function

$$z=f(x_1,\ x_2,\ x_3)\ .$$

We shall assume first that pairs of the variates, say of x_2's and x_3's, are normally distributed. Then by what has just been demonstrated about the form of the correlation surface in three dimensions, the expression

$$(42) \qquad \frac{1}{2\pi\sigma_2\sigma_3[1-r_{23}^2]^{1/2}} e^{-\frac{1}{2(1-r_{23}^2)}\left(\frac{x_2^2}{\sigma_2^2}+\frac{x_2^3}{\sigma_2^2}-2r_{23}\frac{x_2}{\sigma_2}\frac{x_3}{\sigma_3}\right)} dx_2\,dx_3$$

is, to within infinitesimals of higher order, the probability that a point (x_2, x_3) taken at random lies within the area dx_2dx_3. We next assume that the regression of x_1 on x_2 and x_3 is linear, and that each array of x_1's corresponding to an assigned (x_2, x_3) is a normal distribution with standard deviation

$$\sigma_{1.23}=\sigma_1(R/R_{11})^{1/2}$$

given by (32).

Then in the notation of multiple correlation, the probability that a variate taken at random in an assigned (x_2, x_3)-array of x_1's will lie in dx_1 is given, to within infinitesimals of higher order, by

$$(43)\quad \begin{cases} \dfrac{(R_{11})^{1/2}}{\sigma_1(2\pi R)^{1/2}}\, e^{-\frac{R_{11}}{2R\sigma_1^2}\left(x_1+\sigma_1\frac{R_{12}}{R_{11}}\frac{x_2}{\sigma_2}+\sigma_1\frac{R_{13}}{R_{11}}\frac{x_3}{\sigma_3}\right)^2}\, dx_1 \\ \qquad =\dfrac{(R_{11})^{1/2}}{\sigma_1(2\pi R)^{1/2}}\, e^{-\frac{1}{2RR_{11}}\left(R_{11}\frac{x_1}{\sigma_1}+R_{12}\frac{x_2}{\sigma_2}+R_{13}\frac{x_3}{\sigma_3}\right)^2}\, dx_1\,. \end{cases}$$

Then the probability that a point (x_1, x_2, x_3) taken at random will lie in the volume $dx_1dx_2dx_3$ is given, to within infinitesimals of higher order, by the product of (42) and (43). This gives, after some simplification, for the probability in question, $z\, dx_1dx_2dx_3$, where

$$(44)\qquad z=\frac{1}{(2\pi)^{3/2}R^{1/2}\sigma_1\sigma_2\sigma_3}\, e^{-\frac{1}{2}\phi},$$

and

$$\phi=\frac{1}{R}\left(R_{11}\frac{x_1^2}{\sigma_1^2}+R_{22}\frac{x_2^2}{\sigma_2^2}+R_{33}\frac{x_3^2}{\sigma_3^2}+2R_{12}\frac{x_1x_2}{\sigma_1\sigma_2}+2R_{13}\frac{x_1x_3}{\sigma_1\sigma_3}+2R_{23}\frac{x_2x_3}{\sigma_2\sigma_3}\right).$$

38. Certain properties of normally correlated distributions. The equal-frequency curves obtained by making z take constant values in equation (41) are an infinite system of homothetic ellipses, any one of which has an equation of the form

$$\frac{x^2}{\sigma_x^2}+\frac{y^2}{\sigma_y^2}-2r\,\frac{x}{\sigma_x}\,\frac{y}{\sigma_y}=\lambda^2 .$$

The area of the ellipse is

$$\frac{\pi\lambda^2\sigma_x\sigma_y}{(1-r^2)^{1/2}}$$

and the semiaxes are given by $a=k\lambda$ and $b=k'\lambda$, where k and k' are functions of σ_x, σ_y, and r. The probability that a point (x, y) taken at random will fall within any ellipse obtained by assigning λ is given by

$$(45)\qquad \frac{2\pi\sigma_x\sigma_y}{2\pi\sigma_x\sigma_y(1-r^2)}\int_0^\lambda e^{-\frac{1}{2(1-r^2)}\lambda^2}\lambda\, d\lambda=1-e^{-\frac{\lambda^2}{2(1-r^2)}} .$$

Attention has often been called to the equal frequency ellipse known as the "probable" ellipse. The *probable ellipse* may be defined as that ellipse of the system such that the probability is 1/2 that a point (x,y) of the scatter-diagram (see Fig. 18, p. 109) lies within it. This means by (45) that

$$e^{-\frac{\lambda^2}{2(1-r)^2}}=\tfrac{1}{2}\,,\qquad \lambda^2=1.3863\,(1-r^2) .$$

From (45) it follows that $[\lambda/(1-r^2)]\,e^{-\frac{\lambda^2}{2(1-r^2)}}\,\Delta\lambda$ gives, to within infinitesimals of higher order, the probability that a point (x, y) taken at random will fall in a small ring obtained by taking values of λ in $\Delta\lambda$.

We may determine the ellipse[43] along which, for a given small ring $\Delta\lambda$, we should expect more points (x, y) than along any other ellipse of the system. For a constant

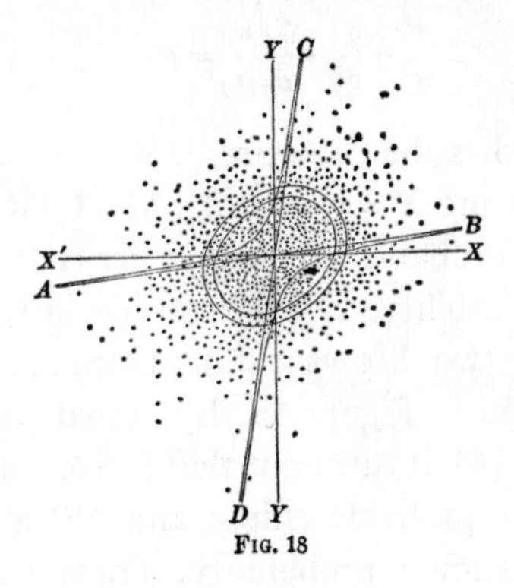

FIG. 18

$\Delta\lambda$, the probability is a maximum when $\lambda^2 = 1 - r^2$. Hence, what may be called the ellipse of maximum probability is

$$\frac{x^2}{\sigma_x^2} + \frac{y^2}{\sigma_y^2} - \frac{2rxy}{\sigma_x \sigma_y} = 1 - r^2 .$$

To illustrate the meaning of this ellipse, we may say that in Bertrand's illustration of shooting a thousand shots at a target, the probability is greater that a shot will strike along this ellipse than along any other ellipse of the system. It is an interesting fact that the ellipse of maximum probability is identical with the orthogonal projection of parabolic points of the correlation surface on the plane of distribution. To prove this theorem, we simply find the locus of parabolic points on the surface (41) by means of the well-known condition

$$\frac{\partial^2 z}{\partial x^2} \cdot \frac{\partial^2 z}{\partial y^2} = \left(\frac{\partial^2 z}{\partial x \partial y}\right)^2 .$$

This gives

$$\frac{x^2}{\sigma_x^2} + \frac{y^2}{\sigma_y^2} - \frac{2rxy}{\sigma_x \sigma_y} = 1 - r^2 ,$$

which establishes the theorem.

By comparing $\lambda^2 = 1 - r^2$ with $\lambda^2 = 1.3863\ (1 - r^2)$, we note that the probable ellipse is larger than the ellipse of maximum probability. For the statures of 1,078 husbands and wives, the two ellipses just discussed are shown on the scatter-diagram in Figure 18. By actual count from the drawing (Fig. 18), it turns out that 536 of the 1,078 points are within the probable ellipse and 412 are within the ellipse of maximum probability. These numbers differ from the theoretical values by amounts well within what should be expected as chance fluctuations.

Another interesting problem in connection with the correlation surface relates to the determination of the locus along which the frequency or density of points on the plane of distribution (scatter-diagram) bears a simple relation to the corresponding density under independence. Thus, we seek the curve along which dots of the scatter-diagram are k times as frequent as they would be under independence where k is a constant. Equating z in (41) to k times the corresponding value of z when $r=0$ in (41), we obtain after slight simplification the hyperbola (Fig. 18)

$$\text{(46)} \qquad \frac{x^2}{\sigma_x^2} + \frac{y^2}{\sigma_y^2} - \frac{2}{r} \frac{xy}{\sigma_x \sigma_y} = \frac{(1-r^2)}{-r^2} \log\ (k^2 - k^2 r^2) .$$

Karl Pearson dealt[44] with this curve for $k=1$. That is, he considered the locus along which the density of points of the scatter-diagram is the same as it would have been under independence. The fact that the density of distribution at the centroid in (41) is $1/(1-r^2)^{1/2}$ times as much as it would be under independence naturally suggests the study of the locus of all points for which $k=1/(1-r^2)^{1/2}$ in (46). It turns out that in this case the hyperbola degenerates into straight lines

$$y=\frac{\sigma_y}{r}\frac{x}{\sigma_x}\left[1\pm(1-r^2)^{1/2}\right].$$

These lines are shown as lines AB and CD on Figure 18. They separate the plane of distribution into four compartments such that one-fourth is the probability that a pair of values (x,y) taken at random will give a point falling into any prescribed one of these compartments.

Although no further discussion of the properties of normal correlation surfaces will be attempted in this monograph, certain properties analogous to those mentioned for the surface in three dimensions would follow rather readily in the case of the surfaces in higher dimensions. Thus the system of ellipsoids of equal frequencies has been studied to some extent.[45] In a paper by James McMahon,[46] the connection between the geometry of the hypersphere and the theory of normal frequency functions of n variables is established by linearly transforming the hyperellipsoids of equal frequency into a family of hyperspherical surfaces, and by applying the formulas of hyperspherical goniometry to obtain theorems in multiple and partial correlations.

39. **Remarks on further methods of characterizing correlation.** In bringing to a conclusion our discussion of correlation, it may be of interest to point out a few of the limitations and omissions in our treatment, and to give certain references that would facilitate further reading.

We have not even touched on the methods of dealing with correlation of characters which do not seem to admit of exact measurement, but admit of classification; for example, eye color, hair color, and temperament may be regarded as such characters. Such characters are sometimes called qualitative characters to distinguish them from quantitative characters. The correlation between two such characters has been dealt with in some cases by the method of tetrachoric[47] correlation, in other cases by the method of contingency,[44] and by the method of correlation in ranks[48] in cases where the items are ordered but not measured. We have not touched on the methods of dealing with correlation[49] in time series—a subject of much importance in the methodology of economic statistics. The methods and theories of connection and concordance of Gini[50] for dealing with correlation have been omitted. No discussion has been given of the fundamental work of Bachelier[51] on correlation theory in his treatment of continuous probabilities of two or more variables. Our discussion of frequency surfaces in §37 is limited to normal correlation surfaces. The way is, however, fairly clear for the extension[24] of the Gram-Charlier system of representation to distributions of two or more variables which are not normally distributed.

While great difficulties have been encountered in the past thirty years in attempts to pass naturally from the Pearson system of generalized frequency curves to analogous surfaces for the characterization of the distribution of two correlated variables, it is of considerable interest to remark that substantial progress has been made recently on the solution of this problem by Narumi,[52] Pearson,[53] and Camp.[54]

Although the many omissions make it fairly obvious that our discussion is not at all complete, it is hoped that enough has been said about the theory of correlation to indicate that this theory may be properly considered as constituting an extensive branch in the methodology of science that should be further improved and extended.

CHAPTER V

RANDOM SAMPLING FLUCTUATIONS

40. Introduction. In Chapter II we have dealt to some extent with the effects of random sampling fluctuations on relative frequencies. But it is fairly obvious that the interest of the statistician in the effects of sampling fluctuations extends far beyond the fluctuations in relative frequencies. To illustrate, suppose we calculate any statistical measure such as an arithmetic mean, median, standard deviation, correlation coefficient, or parameter of a frequency function from the actual frequencies given by a sample of data. If we need then either to form a judgment as to the stability of such results from sample to sample or to use the results in drawing inferences about the sampled population, the common-sense process of induction involved is much aided by a knowledge of the general order of magnitude of the sampling discrepancies which may reasonably be expected because of the limited size of the sample from which we have calculated our statistical measures.

We may very easily illustrate the nature of the more common problems of sampling by considering the determination of certain characteristics of a race of men. For example, suppose we wish to describe any character such as height, weight, or other measurable attributes among the white males age 30 in the race. We should almost surely attempt merely to construct our science on the basis of results obtained from the sample. Then the question arises: What is an adequate sample for a particular purpose? The theory of sampling throws some light on this question. The development of the elements of a theory of sampling fluctuations in various averages, coefficients, and parameters is thus of fundamental importance in regarding the results obtained from a sample as approximate representatives of the results that would be obtained if the whole indefinitely large population were taken.

One of the difficult and practical questions involved in making statistical inquiries by sample relates to the invention of satisfactory devices for obtaining a random sample at the source of material. A result obtained from a sample unless taken with great care may diverge significantly from the true value characteristic of the sampled population. For example, the writer had an experience in attempting to pick up a thousand ears of Indian corn at random with respect to size of ears. It soon appeared fairly obvious that instinctively one tended to make "runs" on ears of approximately the same size. The sample would probably not be taken at random when thus drawn. Such systematic divergence from conditions necessary for obtaining a random sample is assumed to be eliminated before the results that follow from the theory of random sampling fluctuations are applicable. In the practical applications of sampling theory, it is thus important to remember that the conditions for random sampling at the source of data are not always easily fulfilled. In fact, it seems important in certain investigations to devise special schemes for obtaining a random sample. For example, we may sometimes improve the conditions for drawing a random sample of individuals by the use of a ball or card bearing the number of each individual of a much larger aggregate than the sample we propose to measure and by then drawing the sample by lot from such a collection of balls or cards after they have been thoroughly mixed. Even with urn schemata containing white and black balls thoroughly mixed, it must be assumed further that one kind of balls is not more slippery than another if slippery balls evade being drawn. The appropriate devices for obtaining a random sample depend almost entirely on the nature of the particular field of inquiry, and we shall in the following discussion simply assume that random samples can be drawn.

In an inquiry by sample, the following fundamental question comes up very naturally about any result, say a mean value $\bar{x}$, to be obtained from a sample of s individuals: What is the probability that $\bar{x}$ will deviate not more numerically than an assigned positive number δ from the corresponding unknown true value $\bar{\bar{x}}$ that would be given by using an unlimited supply of the material from which the s variates are drawn? This question presents difficulties. An ideal answer is not available, but valuable estimates of the probability called for in this question may be made under certain conditions by a procedure which involves finding the standard deviation of random sampling deviations.

For the unknown true value $\bar{\bar{x}}$ referred to above, continental European writers very generally use the mathematical expectation or the expected value of the variable (cf. § 6). In what follows, we shall to some extent adopt this practice and shall find it convenient to assume the following propositions without taking the space to demonstrate them:

I. The expected value $E[x-E(x)]$ of deviations of a variable from its expected value $E(x)$ is zero.

II. The expected value of the sum of two variables is the sum of their expected values. That is, $E(x+y)=E(x)+E(y)$.

III. The expected value of the product of a constant and a variable is equal to the product of the constant by the expected value of the variable. That is, $E(cx)=cE(x)$.

IV. The expected value of the product xy of corresponding values of two mutually independent variables x and y is equal to the product of their expected values, where we call x and y mutually independent if the law of distribution of each of them remains the same whatever values are assigned to the other.

V. In particular, if x and y are corresponding deviations of two mutually independent variables from their expected values, the expected value of the product xy is zero. It is fairly obvious that V follows from I and IV.

It is convenient in the discussion of random sampling fluctuations to deal with the problem of the distribution of results from samples of equal size. To give a simple example, let us conceive of taking a random sample consisting of 1,000 men of a well-defined race in which some character is measured giving us 1,000 variates. Next, suppose we repeat the process until we have 1,000 such samples of 1,000 men in each sample. Then each of the samples would have its own arithmetic mean, median, mode, standard deviation, moments, and so on. Consider next the 1,000 results of a given kind, say the 1,000 arithmetic means from the samples. They would almost surely differ but slightly from one another in comparison with differences between extreme individual variates. But if the measurements are reasonably accurate the means would differ and form a frequency distribution. This frequency distribution of means would have its own mean (mean of means) and its own standard deviation. We are especially interested in such a standard deviation, for it may be taken as an approximate measure of the variability or dispersion of means obtained from different samples. This standard deviation (standard error) would no doubt be a fairly satisfactory measure of sampling fluctuations for certain purposes.

Although the process of finding mean values from each of a large number of equal samples with a large number of individuals in each sample gives us a useful conception of the problem of sampling errors in mean values, it would ordinarily be a laborious and usually an impractical task because of paucity of available data to carry out such a set of calculations. The statistician ordinarily obtains a result from a sample by calculation, say a mean value $\bar{x}$, and then investigates the standard deviation of such results without taking further samples. That such a treatment of the problem is possible is clearly an important mathematical achievement.

The space available in the present monograph will not permit the derivation of formulas for the standard deviation of sampling errors in many types of averages or parameters. In fact, we shall limit ourselves to presenting only sufficient derivations of such formulas to indicate the nature of the main assumptions and approximations involved in the rationale which supports such formulas, and certain of their interpretations. Preliminary to deriving formulas for standard deviations of sampling errors in certain averages and parameters, we need to find the standard deviation and correlation of errors in class frequencies of any given frequency distribution. For brevity we shall use the expression "standard error" in place of "standard deviation of errors."

41. **Standard error and correlation of errors in class frequencies.** Suppose we obtain from a random sample of a population an observed frequency distribution

$$f_1, f_2, \ldots, f_t, \ldots, f_n$$

with a number f_t of individuals in a class t, and with a total of $f_1+f_2+\ldots+f_n=s$ individuals observed in the sample.

Suppose next that we should obtain a large number of such samples of s observations each taken under the same essential conditions. A class frequency f_t will vary from sample to sample. These values f_t of will form a frequency distribution. We set the problem of expressing the expected value of the square of the standard deviation σ_{f_t} in terms of observed values.

To solve this problem, we may consider that any observation to be made is a trial, and that it is a success to obtain an observation for which the individual falls in the class t. Let p_t be the probability of success in one trial, and $q_t=1-p_t$ be the corresponding probability of failure.

In sets of s trials with a constant probability p_t of obtaining an individual in the class t, we have from page 27 that the square of the standard deviation of f_t in the theoretical distribution is given by

$$\sigma_{f_t}^2=sp_tq_t=sp_t(1-p_t) . \tag{1}$$

In statistical applications, we do not ordinarily know the exact value of p_t, but accept the relative frequency f_t/s as an approximation to p_t if s is large. If we thus accept f_t/s as an approximation to p_t, and substitute $p_t = f_t/s$ in (1), we obtain

$$\sigma_{f_t}^2 = f_t(1 - f_t/s) \tag{2}$$

as an approximate value of the square of σ_{f_t} conveniently expressed in terms of observed frequencies.

The value (2) is regarded as an appropriate approximation to the value of (1) because (1) may be obtained from (2) by replacing the quotient f_t/s by its expected value p_t. It is usually agreed among statisticians, however, that a better approximation to (1) would be an expression which as a whole has the second member of (1) as its expected value. The expected value of the product $f_t(1 - f_t/s)$ is not the product $sp_t(1 - p_t)$ of the expected values of its factors, as we shall see in the next paragraph. It will be found that the second member of the equation

$$\sigma_{f_t}^2 = \frac{s}{s-1} f_t\left(1 - \frac{f_t}{s}\right) \tag{3}$$

has $sp_t(1 - p_t)$ as its expected value, and (3) is therefore regarded as a better approximation than (2) for expressing (1) in terms of observed frequencies. The reason for the advantage of formula (3) over formula (2) is the subject of frequent inquiries by students of statistics, and it is hoped that the discussion here given will contribute to answering such inquiries.

In accordance with the principle just stated it will be seen that the error introduced by replacing $sp_t(1 - p_t)$

by $f_t(1-f_t/s)$ involves not only sampling errors, but also a certain systematic error. Thus, although the expected value of f_t is sp_t (p. 26) and the expected value of $1-f_t/s$ is $1-p_t$, we shall see as stated above that the expected value of the product $f_t(1-f_t/s)$ is not equal to the product $sp_t(1-p_t)$ of the expected values, but is in fact equal to $(s-1)p_t(1-p_t)$. We may prove this by first expressing (1), with the help of the definition of σ_{f_t}, in the form

$$E[(f_t-sp_t)^2]=sp_t(1-p_t)\ , \tag{4}$$

and then applying the last proposition on page 21 which states that the expected value of the square of the variable x is equal to the square of the expected value of x increased by the expected value of the square of the deviations of x from its expected value. Thus, for a variable $x=f_t$ with an expected value sp_t, we write

$$E(f_t^2)=s^2p_t^2+E[(f_t-sp_t)^2]=s^2p_t^2+sp_t(1-p_t)$$

from (4). Further,

$$\left\{\begin{aligned} E[f_t(1-f_t/s)]&=E(f_t)-\frac{1}{s}E(f_t^2)\\ &=sp_t-sp_t^2-p_t(1-p_t)=(s-1)p_t(1-p_t)\ . \end{aligned}\right. \tag{5}$$

By multiplying both members of (5) by $s/(s-1)$, we may write

$$sp_t(1-p_t)=E\left[\frac{s}{s-1}f_t(1-f_t/s)\right]\ .$$

Thus, in approximating to the value $sp_t(1-p_t)$ in the right member of (1) by means of a function of the observed f_t, we note that the function $sf_t(1-f_t/s)/(s-1)$ has the expected value $sp_t(1-p_t)$ which we seek, and that $f_t(1-f_t/s)$ given in the right member of (2) as an approximation to $sp_t(1-p_t)$ contains a systematic error.

In finding standard errors in means, moments, correlation coefficients, and so on, it is important to know the correlation between deviations of frequencies in any two classes. Let δf_t be the deviation of f_t from the theoretical mean or expected value of the class frequency in taking a random sample of s variates. Then since $f_1+f_2+\cdots\cdot+f_t+\cdots\cdot+f_n=s=$ a constant, we have

$$(6) \qquad \delta f_1+\delta f_2+\cdots\cdot+\delta f_t+\cdots\cdot+\delta f_n=0\,.$$

If our sample has given δf_t more than the expected number in the class t, it may reasonably be assumed that a deficiency equal to $-\delta f_t$ will tend to be distributed among the other groups in proportion to their expected relative frequencies.

Now suppose we had a correlation table made of pairs of values of δf_t and $\delta f_{t'}$, obtained from a large number of samples. Consider the array in which δf_t has a fixed value By (6), for each sample,

$$-\delta f_t=\delta f_1+\delta f_2+\cdots\cdot+\delta f_{t-1}+\delta f_{t+1}+\cdots\cdot+\delta f_n\,.$$

Assume that the amount of frequency in the left member of this equality is distributed to terms of the right member in such proportion that, for a fixed δf_t, the mean value of $\delta f_{t'}$ is

$$(7) \qquad -\delta f_t\cdot\frac{sp_{t'}}{s-sp_t}=-\delta f_t\cdot\frac{p_{t'}}{1-p_t}\,.$$

This gives the mean of the array under consideration.

It is fairly obvious that the correlation coefficient

$$r = \frac{1}{N} \sum_y \sum_x \frac{x}{\sigma_x} \frac{y}{\sigma_y}$$

of N pairs of deviations x and y from mean values is equal to

$$\frac{1}{N} \sum_x \frac{x}{\sigma_x} \frac{\bar{y}_x N_x}{\sigma_y},$$

where $\bar{y}_x$ is the mean of the x-array of y's, and N_x is the number in the array. Then

$$r\sigma_x\sigma_y = \text{mean value of } x\bar{y}_x = \frac{1}{N} \sum_x x\bar{y}_x N_x .$$

By attaching this meaning to the correlation coefficient $r_{f_t f_{t'}}$ of f_t and $f_{t'}$ and using (7) for the mean of the array, we have

$$r_{f_t f_{t'}} \sigma_{f_t} \sigma_{f_{t'}} = \text{mean value of } -\delta f_t \cdot \frac{\delta f_t p_{t'}}{1-p_t}$$

$$= \frac{-p_{t'}}{1-p_t} (\text{mean value of } \delta f_t^2) = \frac{-p_{t'}}{1-p_t} \sigma_{ft}^2$$

$$= -sp_t p_{t'} \text{ from (1)} \tag{8}$$

$$= -\frac{f_t f_{t'}}{s} \tag{9}$$

as a first approximation.

A systematic error is involved in replacing $sp_t p_{t'}$ by $f_t f_{t'}/s$ on account of the correlation between f_t and $f_{t'}$.

To deal with the effect of this correlation, we may first write (3), page 83, in the form

$$\frac{1}{N}\sum_{i=1}^{i=N} x_i y_i = \bar{x}\bar{y} + r\sigma_x\sigma_y \,.$$

If we are dealing with a population or theoretical distribution rather than with a sample, this formula gives us the proposition that the expected value of the product, $x_i y_i$, of pairs of variables is equal to the product, $\bar{x}\bar{y}$, of their expected values increased by the product, $r\sigma_x\sigma_y$, of the correlation coefficient and the two standard deviations.

To apply this proposition when $x_i = f_t$ and $y_i = f_{t'}$, we note from (8) that, for the population, $r\sigma_x\sigma_y = -sp_t p_{t'}$, and recall that $E(f_t) = sp_t$ and $E(f_{t'}) = sp_{t'}$. Then the proposition stated above gives us

$$E(f_t f_{t'}) = s^2 p_t p_{t'} - sp_t p_{t'} \,,$$

$$\text{(10)} \qquad E(f_t f_{t'}/s) = \frac{1}{s} E(f_t f_{t'}) = (s-1) p_t p_{t'} \,.$$

To obtain the right member of (8) as accurately as possible in terms of the observed f_t and f_t, we multiply both members of (10) by $s/(s-1)$ and then note that $f_t f_{t'}/(s-1)$ has the expected value $sp_t p_{t'}$. In the right member of (9), the value $f_t f_{t'}/s$ used as an approximation to $sp_t p_{t'}$ thus contains a certain systematic error. To eliminate the systematic error from (9), we write

$$\text{(11)} \qquad r_{f_t f_{t'}} \sigma_{f_t} \sigma_{f_{t'}} = -\frac{f_t f_{t'}}{s-1}$$

in place of (9) as a second approximation to (8).

42. Remarks on the assumptions involved in the derivation of standard errors. The three outstanding assumptions that should probably be emphasized in considering the validity and the limitations of the results (2) and (9) are (*a*) that the probability that a variate taken at random will fall into any assigned class remains constant, (*b*) that the number s is so large that we obtain certain valuable approximations by using the relative frequency f_t/s in place of the probability p_t that a variate taken at random will fall into the class t, and (*c*) that any sampling deviation δf_t from the expected value of a class frequency is accompanied by an apportionment of $-\delta f_t$ to other class frequencies in amounts proportional to the expected values of such other class frequencies. Our use of assumption (*b*) involves more than is apparent on the surface, because in its use we not only replace a single isolated probability p_t by a corresponding relative frequency f_t/s, but we further assume the liberty of using certain functions of the relative frequencies in place of these functions of the corresponding probabilities or expected values. This procedure may lead to certain systematic errors in addition to the sampling errors. For example, we have, in obtaining (2), used the function $f_t(1-f_t/s)$ of f_t/s in place of the same function $sp_t(1-p_t)$ of the expected value p_t, and have by this procedure tended to underestimate the expected value when s is finite. That is, $sf_t(1-f_t/s)/(s-1)$ and not $f_t(1-f_t/s)$ is our best estimate of the expected value. However, when s becomes large, $f_t(1-f_t/s)$ is a valuable first approximation to the expected value.

The rule that the expected value of a function may be taken as approximately equal to the function of the expected value has been much used by statisticians in a rather loose and uncritical manner. A critical study of the application and limitations of this rule was published by Bohlmann[55] in 1913. While it is beyond the scope of this monograph to enter upon a general discussion of Bohlmann's conclusions, it is of special interest for our purpose that the application of the rule leads at least to first approximations when the functions in question are algebraic functions. Although it may seem that we have in the derivation of (2) and (9) taken the liberty to substitute relative frequencies rather freely in place of the probabilities required in an exact theory, this procedure may be extended to any algebraic functions when the number s is very large, with the expectation of obtaining useful approximations. Since certain derivations which follow make use of (2) and (9), the resulting formulas involve the weaknesses and limitations of the above assumptions.

43. Standard error in the arithmetic mean and in a *q*th moment coefficient about a fixed point. For the arithmetic mean of s observed values of a variable x we write

$$\bar{x}=\frac{1}{s}\sum_{t=1}^{t=n} x_t f_t ,$$

where f_t is the class frequency of x_t.

Suppose the s values constitute a random sample of observations on the variable x. Suppose further that we continue taking observations on x until we have a very large number of random samples each consisting of s observed values. Then assume that there exists an expected value of each f_t about which the observed f_t's exhibit dispersion, and that corresponding to these expected values there exists a theoretical mean value $\bar{\bar{x}}$ of $\bar{x}$ about which the $\bar{x}$'s calculated from samples of s exhibit dispersion. Using δf and $\delta\bar{x}$ to denote deviations in any sample from the expected values of f and $\bar{x}$, respectively, we write

$$s\delta\bar{x} = \Sigma x_t \delta f_t \,,$$
$$s^2(\delta\bar{x})^2 = \sum(x_t^2 \overline{\delta f}_t) + 2\sum{}'(x_t x_{t'} \delta f_t \delta f_{t'}) \,,$$

where the sum $\sum$ extends from $t=1$ to $t=n$, and $\sum{}'$ is the sum for all values of t and t' for which $t \neq t'$.

Next, sum both members of this equality for all samples and divide by the number of samples. This gives in the notation for standard deviations (p. 119) and for the correlation coefficient (p. 123),

$$s^2\sigma_{\bar{x}}^2 = \sum(x_t^2\sigma_{f_t}^2) + 2\sum{}'(x_t x_{t'} \sigma_{f_t} \sigma_{f_{t'}} r_{f_t f_{t'}}) \,.$$

By using (1) and (8), we have

$$\begin{aligned} s\sigma_{\bar{x}}^2 &= \sum(x_t^2 p_t) - \sum(x_t^2 p_t^2) - 2\sum{}'(x_t x_{t'} p_t p_{t'}) \\ &= \mu_2' - (\Sigma x_t p_t)^2 = \mu_2' - \bar{x}^2 = \sigma^2 \,, \end{aligned}$$

where σ is the standard deviation of the theoretical distribution. Then

$$\sigma_{\bar{x}} = \frac{\sigma}{s^{1/2}} \,. \tag{12}$$

Instead of the σ of the theoretical distribution, we ordinarily use the σ obtained from a sample. To introduce the expected value of σ^2 from the sample, we may, for a first approximation, use (2) and (9) in place of (1) and (8) above, and obtain very simply a form identical with (12).

As a second approximation, we may use (3) and (11) in place of (1) and (8) above, and obtain very simply

$$(13) \qquad \sigma_{\bar{x}}^2 = \frac{\sigma^2}{s-1} \text{ and } \sigma_{\bar{x}} = \frac{\sigma}{(s-1)^{1/2}}$$

where σ is to be obtained from the sample.

The distinction between the expected value of σ^2 from the population and from the sample involves a rather delicate point, but one that has been long recognized in the literature of error theory. The distinction has been rather generally ignored in books on statistics. In numerical problems, the differences in the results of formulas (12) and (13) are negligible when s is large.

The standard deviation (standard error) may well serve as a measure of sampling fluctuations. But custom has not established the direct use of the standard error to any considerable extent. The so-called probable error has come into much more common use than the standard error. The probable error E is sometimes defined very simply as .6745 times the standard error without regard to the nature of the distribution. This definition of the probable error does not impose the condition that the distribution of results obtained on repetition shall necessarily be a normal distribution. But with such a definition of probable error, the real difficulty is not overcome, but merely shifted to the point where we attempt an interpretation of the probable error in terms of the odds in favor of or against an observed result obtained from a sample falling within an assigned deviation of the true value.

Thus, in the derivation of (12) we have obtained, subject to certain important limitations, the standard deviation of means $\bar{x}$ obtained from samples of s about a theoretical mean value $\bar{\bar{x}}$ which may ordinarily be regarded as a sort of a true value of the mean. If the distribution of $\bar{x}$'s obtained from samples about such a true value is assumed to be a normal distribution, we may by the use of the table of the probability integral state at once that the odds are even that an $\bar{x}$ obtained from a sample will differ numerically from the true value by not more than

$$E = .6745 \text{ (standard error)} .$$

It is the assumption of a normal distribution of the means from samples combined with the specification of an even wager that brings the multiplier .6745 into the problem.

We may further expedite the treatment of sampling errors by finding the odds in favor of or against an observed deviation from the true value not exceeding numerically a certain multiple of E, say tE. As t increases to 5, 6, or more, the odds in favor of obtaining a deviation smaller than tE are so large as to make it practically certain that we will obtain such a smaller deviation.

We have discussed briefly the meaning and limitations of probable errors. The most outstanding limitation on the interpretation of probable errors is the requirement of a normal distribution of the statistical constant under consideration. We have to a considerable extent used the arithmetic mean as an illustration, but the same general requirements about the normality of the distribution would clearly apply, whatever the statistical constant.

We shall consider next the standard error in a qth moment coefficient μ'_q about a fixed point. By definition,

$$s\mu'_q = \sum (x_t^q f_t) .$$

For the relation between deviations from theoretical values we have

$$s\delta\mu'_q = \sum(x_t^q \delta f_t) .$$

Then

$$s^2\overline{\delta\mu'^2_q} = \sum\left(x_t^{2q}\overline{\delta f_t^2}\right) + 2\sum{}'(x_t^q x_{t'}^q \delta_{f_t}\delta_{f_{t'}}) .$$

Sum both members of this equality for a large number of samples N and divide by N. This gives in the notation for standard deviations (p. 119) and for the correlation coefficient (p. 123)

$$s^2\sigma^2_{\mu'_q} = \sum(x_t^{2q}\sigma^2_{f_t}) + 2\sum{}'(x_t^q x_{t'}^q \sigma_{f_t}\sigma_{f_{t'}} r_{f_t f_{t'}}) .$$

Using (1) and (8), we have

$$\begin{aligned} s\sigma^2_{\mu'_q} &= \sum(x_t^{2q} p_t) - \sum(x_t^{2q} p_t^2) - 2\sum{}'(x_t^q x_{t'}^q p_t p_{t'}) \\ &= \mu'_{2q} - (\sum x_t^q p_t)^2 \\ &= \mu'_{2q} - \mu'^2_q . \end{aligned}$$

Then

$$(14) \qquad \sigma_{\mu'} = \left[\frac{\mu'_{2q} - \mu'^2_q}{s}\right]^{1/2} ,$$

where the moments in the right-hand member relate to the theoretical distribution. By methods analogous to those used in the case of the arithmetic mean (pp. 127–28), we may pass to moments which relate to the sample.

The probable error of μ'_q is then $E = .6745\sigma_{\mu'_q}$, and the usual interpretation of such a probable error by means of odds in favor of or against deviations less than a multiple of E is again dependent on the assumption that the qth moments μ'_q found from repeated trials form a normal distribution.

44. Standard error of the qth moment μ_q about a mean. In considering the problem of the standard error of a moment about a mean, it is important to recognize the difference between the mean of the population and a mean obtained from a sample.

For simplicity, we shall consider the problem of the standard error in a qth moment about the mean of the population when we take samples of s variates as in § 43. The mean of the population is a fixed point about which we take the qth moment of each sample of s variates. Then if we follow the usual plan of dropping the primes from the μ's to denote moments about a mean, we write from (14)

$$\sigma^2_{\mu_q} = (\mu_{2q} - \mu^2_q)/s$$

for the square of the standard error of μ_q in terms of moments of the theoretical distribution.

In particular, we have for the standard error of the second moment

$$\sigma^2_{\mu_2} = (\mu_4 - \mu_2^2)/s \, .$$

When the distribution is normal,

$$\mu_4 = 3\mu_2^2 \, , \qquad \text{and} \qquad \sigma^2_{\mu_2} = 2\mu_2^2/s \, .$$

Since $\sigma = (\mu_2)^{1/2}$, we have

$$\delta\sigma = \frac{\delta\mu_2}{2(\mu_2)^{1/2}}$$

nearly.

Square each member, sum for all samples, and divide by the number of samples. This gives

$$\sigma_\sigma^2 = \frac{\sigma_{\mu_2}^2}{4\sigma^2} = \frac{2\mu_2^2}{4s\sigma^2} = \frac{\sigma^2}{2s} \qquad \text{or} \qquad \sigma_\sigma = \sigma/(2s)^{1/2}.$$

Hence, the probable error in approximating to the standard deviation σ of the population by the standard deviation from a sample of s variates is given approximately by

$$.6745\sigma_\sigma = .6745\sigma/(2s)^{1/2}.$$

To avoid misunderstanding, it should perhaps be emphasized that we have throughout this section restricted our discussion to the qth moment about the mean of the population. The problem of dealing with the standard error in the qth moment about the mean of a sample offers additional difficulties because such a mean varies from sample to sample. A problem arises from the correlation of errors in the means and in the corresponding moments. Further problems arise in considering the closeness of certain approximations, especially when the moments are of fairly high order, that is, when q is large. We shall simply state without demonstration that the square of the standard error in the qth moment about the mean of a sample is given by

$$\frac{(\mu_{2q} - \mu_q{}^2 - 2q\mu_{q+1}\mu_{q-1} + q^2\sigma^2\mu_{q-1})}{s}$$

as a first approximation. For $q=2$, this expression becomes $(\mu_4 - \mu_2{}^2)/s$. For $q=4$, it becomes $(\mu_8 - \mu_4{}^2)/s$ in the case of a normal distribution. These expressions for the special cases $q=2$ and $q=4$ are the same as for the moments about a fixed point.

45. Remarks on the standard errors of various statistical constants. We have shown a method of derivation of the standard errors in certain statistical constants (the mean, the qth moment about a fixed point), and in particular the derivation of probable error of the mean. Our main purpose has been to indicate briefly the nature of the assumptions involved in the derivation of the most common probable-error formulas. The next step would very naturally consist in finding the correlations of errors in two moments. Following this, we could deal with the general problem of standard errors in parameters of frequency functions of one variable on the assumption that the parameters may be expressed in terms of moment coefficients. Thus, let

$$y=f(x, c_1, c_2, \ldots .)$$

be any frequency curve, where any parameter

$$c_i=\phi(\bar{x}, \mu_2, \mu_3, \ldots . , \mu_q, \ldots .)$$

is a function of the mean and of moments about the mean.

Suppose that this relation is such that we may express δc_i in terms of $\delta\bar{x}$, $\delta\mu_2$, $\delta\mu_3$, ... , at least approximately by differentiation of the function ϕ. If we then square δc_i, sum, and divide by the number of samples, we obtain an approximation to the square of the standard error in c_i.

While, in a general way, this method may be described as a straightforward procedure, the derivation of useful formulas is likely to involve rather laborious algebraic details. Moreover, considerable difficulty may arise in estimating the errors involved in the approximate results.

The difficulties of estimating the magnitude of the

errors involved are likely to be much increased when the statistical constant, for example, a correlation coefficient, is a function not merely of the moments of the separate variables, but also of the product moments of two variables.

In concluding these remarks on standard errors of statistical parameters obtained from moments of observations, it may be of interest to point out that the characterization of the sampling fluctuations in such parameters may be extended and refined by the use of higher-order moments of the errors in the parameters. B. H. Camp has shown that the use of moments of order higher than two may very naturally be accompanied by the use of a certain number of terms of Gram-Charlier series as a distribution function.[56]

46. Standard error of the median. Thus far in our discussion of standard errors and probable errors, we have assumed that the statistical constants or characteristics of the frequency function are given as functions of the moments. There are, however, useful characteristics such as a median, a quartile, a decile, and a percentile of a distribution which were not ordinarily given as functions of moments. Such a characteristic number used in the description of a distribution is ordinarily calculated from its definition, which specifies that its value is such that a certain fractional part of the total frequency is on either side of the value in question. For example, a median m of a given distribution is ordinarily calculated from the definition that variates above and below m are to be equally frequent. Similarly, a fourth decile D_4 is calculated from the definition that four-tenths of the frequency is to be below D_4. We are thus concerned with the sampling fluctuations of the bounds of the interval which includes an assigned proportion of the frequency.

To illustrate further, let us consider the standard error in the median m of samples of N of a variable x distributed in accord with a continuous law of frequency given by $y=f(x)$. We assume that there exists a certain ideal median value M of the population of which we have a sample of N and that by definition of the median $1/2$ is then the probability that a variate taken at random falls above (or below) M. We may then write that in any sample of N variates taken at random from the indefinitely large set, the number above M is $N/2+d$. That is, the median m of the sample is at a distance $\delta x=\delta m$ from M. When y has a value corresponding to a value of x in the interval δm, we may write

$$y\delta m=d$$

to within infinitesimals of higher order.

Such an equation connects the change δm in the median of the sample from the theoretical M with the sampling deviation d of the frequency above M. Then

$$\delta m=\frac{d}{y} \quad \text{and} \quad \sigma_m^2=\frac{1}{y^2}\sigma_d^2 .$$

But, from (1), page 119,

$$\sigma_d^2=Npq=\frac{N}{4} . \qquad \text{Hence} \qquad \sigma_m=\frac{N^{1/2}}{2y}.$$

If we have a normal distribution

$$y=f(x)=\frac{N}{\sigma(2\pi)^{1/2}}\,e^{-(x-M)^2/2\sigma^2},$$

the value of y at the median is given by

$$y=\frac{N}{\sigma(2\pi)^{1/2}}=.39894\,\frac{N}{\sigma}\,,$$

and the standard error in the median found from ranks is

$$\sigma_m=\frac{1.2533\sigma}{N^{1/2}}\,. \tag{15}$$

Although the theoretical values of the median and of the arithmetic mean are equal in a normal distribution, the median found from a sample by ranking has a sampling error 1.2533 times as large as the arithmetic mean obtained as a first moment from the same sample.

47. Standard deviation of the sum of independent variables. In sampling problems, it is often found useful to know the expected value of the square of the standard deviation of the sum $Y=X_1+X_2+\cdots+X_s$ of s mutually independent variables when we have given the standard deviations $\sigma_1, \sigma_2, \ldots, \sigma_s$ of each variable in the population to which it belongs.

Assuming that the given deviations are measured from the theoretical or expected values for the populations, we consider deviations $x_i=X_i-E(X_i)$, and write the deviation of the sum

$$y=x_1+x_2+\cdots+x_s\,.$$

Square both sides, sum for the number of samples N, and divide by N. Then we have

$$\frac{1}{N}\sum y^2=\frac{1}{N}\sum x_1^2+\frac{1}{N}\sum x_2^2+\cdots+\frac{1}{N}\sum x_s^2$$

$$+\frac{2}{N}\sum x_1x_2+\frac{2}{N}\sum x_1x_3+\cdot$$

If we pass to expected values, and let $\sigma_1^2, \sigma_2^2, \ldots, \sigma_s^2$ denote the squares of standard deviations of the several variables and σ_y^2 that of their sum in the populations, we have

$$\sigma_y^2 = \sigma_1^2 + \sigma_2^2 + \cdots + \sigma_s^2 , \tag{16}$$

the product terms vanishing by V, page 117.

It is a matter of some interest to note how the expected value just found differs from the expected value of the sum of squares of the s deviations of $x_1, x_2, \ldots, x_s$ from their mean

$$\frac{1}{s}\sum_{i=1}^{s} x_i$$

obtained from a sample. If we let

$$x_i' = x_i - \frac{1}{s}\sum_{i=1}^{s} x_i , \tag{17}$$

we are to find $E(x_1'^2 + x_2'^2 + \cdots + x_s'^2)$, in terms of $E(x_i^2) = \sigma_i^2$ $(i = 1, 2, \ldots, s)$. From (17) we may write

$$x_1' = \frac{s-1}{s}x_1 - \frac{x_2}{s} - \cdots - \frac{x_s}{s} ,$$

$$x_2' = \frac{s-1}{s}x_2 - \frac{x_1}{s} - \cdots - \frac{x_s}{s} ,$$

$$x_s' = \frac{s-1}{s}x_s - \frac{x_1}{s} - \cdots - \frac{x_{s-1}}{s} .$$

Then for $i \neq j$ we have

$$x_1'^2 + \cdots + x_s'^2 = \frac{s-1}{s}(x_1^2 + \cdots + x_s^2) - \frac{2}{s}\sum x_i x_j .$$

Hence, passing to expected values, using V, page 117,

$$(18)\qquad E(x_1'^2+\cdots+x_s'^2)=\frac{s-1}{s}(\sigma_1^2+\cdots+\sigma_s^2)\,.$$

48. **Remarks on recent progress with sampling errors of certain averages obtained from small samples.** In the development of the theory of sampling, the assumption has usually been made that the sample contains a large number of individuals, thus leading to the expectation that the replacement of probabilities by corresponding relative frequencies will give a valuable approximation. But the lower bound of large numbers has remained poorly defined in this connection. For example, certain probable-error formulas have been applied to as few as ten observations.

Beginning with a paper by Student[57] in 1908 there have been important experimental and theoretical results obtained on the distribution of arithmetic means, standard deviations, and correlation coefficients obtained from small samples.

In 1915, Karl Pearson[58] took an important step in advance by obtaining the curve

$$(19)\qquad y=y_0x^{n-2}e^{-\frac{nx^2}{2\sigma^2}}$$

for the distribution of the standard deviations of samples of n variates from an infinite population distributed in accord with the normal curve.

By finding the moments μ_2, μ_3, and μ_4 of this theoretical distribution, and then tabulating the corresponding

$$\beta_1=\frac{\mu_3^2}{\mu_2^3}\,,\qquad \beta_2=\frac{\mu_4}{\mu_2^2}\,,$$

and the skewness of the curve (19) for integral values of n from 4 to 100, and making use of the fact that $\beta_1=0$, $\beta_2=3$, and sk (skewness) $=0$ are necessary conditions for a normal distribution, Pearson shows experimentally that the distribution of standard deviations given by (19) approaches practically a normal distribution as n increases. In this experiment, the necessary conditions $\beta_1=0$, $\beta_2=3$, and $sk=0$ are assumed to be sufficient for practical approach to a normal distribution.

From this table of values, Pearson concludes that for samples of 50 the usual theory of probable error of the standard deviation holds satisfactorily, and that to apply it to samples of 25 would not lead to any error of importance in the majority of statistical problems. On the other hand, if a small sample, $n<20$ say, of a population be taken, the value of the standard deviation found from the sample tends to be less than the standard deviation of the population.

In a paper published in 1915, R. A. Fisher[59] dealt with the frequency distribution of the correlation coefficient r derived from samples of n pairs each taken at random from an infinite population distributed in accord with the normal correlation surface (p. 104), where ρ is the correlation coefficient. The frequency function $y_n=f_n(r)$ given by Fisher for the distribution of r was such that the investigation of its approach to a normal curve as n increases seemed to require special methods for computing the ordinates and moments. Such special methods were given in a joint memoir[60] by H. E. Soper, A. W. Young, B. M. Cave, A. Lee, and Karl Pearson. The values of β_1 and β_2 were computed for these distributions to study the approach to the normal curve.

With respect to the approach of these distributions to the normal form with increasing values of n, it is found that the necessary conditions $\beta_1=0$, $\beta_2=3$ for a normal distribution are not well fulfilled for samples of 25 or even 50, whatever the value of ρ. For samples of 100, the approach to the conditions $\beta_1=0$, $\beta_2=3$ is fair for low values of ρ, but for large values of ρ, say $\rho>.5$, there is considerable deviation of β_1 from 0, and of β_2 from 3. For samples of 400, on the whole, the approach to the necessary conditions $\beta_1=0$, $\beta_2=3$ is close, but there is quite a sensible deviation from normality when $\rho\geqq.8$. These results give us a striking warning of the dangers in interpreting the ordinary formula for the probable error of r when we have small samples.

As to the limitations on the generality of these results, it should be remembered that the assumption is made, in this theory of the distribution of r from small samples, that we have drawn samples from an infinite population well described by a normal correlation surface, so that the conclusions are not in the strictest sense applicable to distributions not normally distributed. While the results just now described have thrown much light on the distributions of statistical constants calculated from small samples, it is fairly obvious that much remains to be done on this important problem.

49. **The recent generalizations of the Bienaymé-Tchebycheff criterion.** Although the use of probable errors for judging of the general order of magnitude of the numerical values of sampling deviations is a great aid to common-sense judgment, it must surely be granted that we are much hampered in drawing certain inferences depending on probable errors because of the limitation that the interpretation of the probable error of a statistical con-

stant is to some extent dependent in any particular case on the normality of the distribution of such constants obtained from samples, and because of the lack of knowledge as to the nature of the distribution.

Any theory that would deal effectively with the problem of finding a criterion for judging of the magnitude of sampling errors with little or no limitation on the nature of the distribution would be a most welcome contribution, especially if the theory could be made of value in dealing with actual statistical data. The Bienaymé-Tchebycheff criterion (p. 29) may be regarded as an important step in the direction of developing such a theory. We have in the Tchebycheff inequality a theorem specifying an upper bound $1/\lambda^2$ for the probability that a datum taken at random will be equal to or greater than λ times the standard deviation without limitation on the nature of the distribution. That is, if $P(\lambda\sigma)$ is the probability that a datum drawn at random from the entire distribution will differ in absolute value from the mean of all values as much as $\lambda\sigma$, then

$$P(\lambda\sigma) \leqq \frac{1}{\lambda^2} . \tag{20}$$

To establish a first generalization of this inequality (cf. p. 29), let us consider a variable x which takes mutually exclusive values $x_1, x_2, \ldots, x_n$ with corresponding probabilities, $p_1, p_2, \ldots, p_n$, where $p_1+p_2+\ldots+p_n=1$.

Let a be any number from which we wish to measure deviations. For the expected values of the moment of order $2s$ about a, we may write

$$\mu'_{2s} = p_1 d_1^{2s} + p_2 d_2^{2s} + \cdots + p_n d_n^{2s} ,$$

where $d_i = x_i - a$.

Let $d', d'', \ldots$, be those deviations $x_i - a$ which are numerically as large as an assigned multiple $\lambda\sigma$ $(\lambda>1)$ of the root-mean-square deviation, and let $p', p'', \ldots$, be the corresponding probabilities. Then we have

$$\mu'_{2s} \geqq p'd'^{2s} + p''d''^{2s} + \cdots .$$

Since $d', d'', \ldots$, are each numerically as large as $\lambda\sigma$, we have

$$\mu'_{2s} \geqq \lambda^{2s}\sigma^{2s}(p' + p'' + \cdots) .$$

If we let $P(\lambda\sigma)$ be the probability that a value of x taken at random will differ from a numerically by as much as $\lambda\sigma$, then $P(\lambda\sigma) = p' + p'' + \ldots$, and

$$\mu'_{2s} \geqq \lambda^{2s}\sigma^{2s}P(\lambda\sigma) .$$

Then

$$P(\lambda\sigma) \leqq \frac{\mu'_{2s}}{\lambda^{2s}\sigma^{2s}} ,$$

and the probability of obtaining a deviation numerically less than $\lambda\sigma$ is greater than

$$1 - \frac{\mu'_{2s}}{\lambda^{2s}\sigma^{2s}} .$$

This generalization of the Tchebycheff inequality is due to Karl Pearson[61] except that he assumed a distribution given by a continuous function with a as the mean x-coordinate of the centroid of frequency area. For this case, we should merely drop the prime from μ'_{2s}, and write

$$P(\lambda\sigma) \leqq \frac{\mu_{2s}}{\lambda^{2s}\sigma^{2s}} . \tag{21}$$

With $s=1$, we obviously have the Tchebycheff inequality as a special case.

It is Pearson's view that, although his inequality is in most cases a closer inequality than that of Tchebycheff, it is usually not close enough to an equality to be of practical assistance in drawing conclusions from statistical data. On the whole, Pearson expresses not only disappointment at the results of the Tchebycheff inequality, but holds that his own generalization still lacks, in general, the degree of approximation which would make the result of real value in important statistical applications. Hence, it is an important problem to obtain closer inequalities. The problem of closer inequalities has been dealt with in recent papers by several mathematicians.[62] Camp, Guldberg, Meidel, and Narumi have succeeded particularly well by placing certain mild restrictions on the nature of the distribution function $F(x)$. The restrictions are of such a nature as to leave the distribution function sufficiently general to be useful in the actual problems of statistics. The main restriction placed on $F(x)$ by Camp is that it is to be a monotonic decreasing function of $|x|$ when $|x| \geqq c\sigma$, $c \geqq 0$. The general effect of this restriction is to exclude distributions which are not represented by decreasing functions of $|x|$ at points more than a certain assigned distance from the origin. We shall now present the main results of Camp without proof.

With the origin so chosen that zero is at the mean, he reaches the generalized inequality

$$P(\lambda\sigma) \leqq \frac{\beta_{2s-2}}{\lambda^{2s}} \frac{\left(\frac{2s}{2s+1}\right)^{2s}}{1+\phi} + \frac{\phi}{1+\phi} P(c\sigma) \,, \tag{22}$$

where

$$\beta_{2s-2}=\frac{\mu_{2s}}{\sigma^{2s}} \quad \text{and} \quad \phi=\frac{\left(\frac{c}{\lambda}\cdot\frac{2s}{2s+1}\right)^{2s}}{(2s+1)\left(\frac{\lambda}{c}-1\right)}.$$

When $c=0$, the formula (22) is Pearson's formula (21) divided by $(1+1/2s)^{2s}$.

The general effect of the work of Camp and Meidel has been to decrease the larger number of the Pearson inequality (21) by roughly 50 per cent. These generalizations seem to have both theoretical and practical value when we have regard for the fact that the results apply to almost any type of distribution that occurs in practical applications. Indeed, it is so satisfying to have only very mild restrictions on the nature of the distribution in judging sampling errors that further progress in extending the cautious limits of sampling fluctuations given by the generalizations of the Tchebycheff inequality would be of fundamental value.

50. **Remarks on the sampling fluctuations of an observed frequency distribution from the underlying theoretical distribution.** If we have fitted a theoretical frequency curve to an observed distribution, or if we know the theoretical frequencies from a priori considerations, the question often arises as to the closeness of fit of theory and observation. In considering this question, a criterion is needed to assist common-sense judgment in testing whether the theoretical curve or distribution fits the observed distribution well or not. It is beyond the scope of the present monograph to deal with the theory underlying such a criterion, but it seems desirable to remark that the fundamental paper on this important problem of

random sampling was contributed by Karl Pearson under the title, "On the criterion that a given system of deviations from the probable in the case of a correlated system of variables is such that it can be reasonably supposed to have arisen from random sampling," *Philosophical Magazine*, Volume 50, Series 5 (1900), pages 157–75.

Closely related to the problem of the closeness of fit of theory and observation is the fundamental problem of establishing a criterion for measuring the probability that two independent distributions of frequency are really random samples of the same population. Pearson published one solution of this problem in *Biometrika*, Volume 8 (1911–12), pages 250–54. The resulting criterion represents an important achievement of mathematical statistics as an aid to common-sense judgment in considering the circumstances surrounding the origin of a random sample of data.

CHAPTER VI

THE LEXIS THEORY

51. Introduction. We have throughout Chapter II assumed a constant probability underlying the frequency ratios obtained from observation. It is fairly obvious that frequency ratios are often found from material in which the underlying probability is not constant. Then the statistician should make use of all available knowledge of the material for appropriate classification into subsets for analysis and comparison. It thus becomes important to consider a set of observations which may be broken into subsets for examination and comparison as to whether the underlying probability seems to be constant from subset to subset. In the separation of a large number of relative frequencies into n subsets according to some appropriate principle of classification, it is useful to make the classification so that the theory of Lexis may be applied. In the theory of Lexis we consider three types of series or distributions characterized by the following properties:

1. The underlying probability p may remain a constant throughout the whole field of observation. Such a series is called a *Bernoulli series*, and has been considered in Chapter II.

2. Suppose next that the probability of an event varies from trial to trial within a set of s trials, but that the several probabilities for one set of s trials are identical to those of every other of n sets of s trials. Then the series is called a *Poisson series*.

3. When the probability of an event is constant from

trial to trial within a set but varies from set to set, the series is called a *Lexis series*.

The theory of Lexis[63] uses these three types as norms for comparison of the dispersions of series which arise in practical problems of statistics. An estimate of the importance of this theory may probably be formed from the facts that Charlier[64] states in his *Vorlesungen über mathematischen Statistik* (1920) that it is the first essential step forward in mathematical statistics since the days of Laplace, and that J. M. Keynes[65] expressed a somewhat similar opinion in his *Treatise on Probability* (1920). These may be somewhat extreme views when we recall the contributions of Poisson, Gauss, Bravais, and Tchebycheff but they at least throw light on the outstanding character of the contribution of Lexis to the theory of dispersion. The characteristic feature of the method of Lexis is that it encourages the analysis of the material by breaking up the whole series into a set of sub-series for examination of the fluctuation of the frequency among the various sub-series. Such a plan of analysis surely has the sanction of common-sense judgment.

In drawing s balls one at a time with replacements from an urn of such constitution that p is the constant probability that a ball to be drawn will be white, we have already established the following results for Bernoulli series:

1. The mathematical expectation of the number of white balls is sp (p. 26).

2. The standard deviation of the theoretical distribution of frequencies is $(spq)^{1/2}$ (p. 27).

3. The standard deviation of the corresponding distribution of relative frequencies is $(pq/s)^{1/2}$ (p. 27).

52. Poisson series. To develop the theory of the Poisson series let s urns,

$$U_1, U_2, \ldots, U_s,$$

contain white and black balls in such relative numbers that

$$p_1, p_2, \ldots, p_s$$

are the probabilities corresponding to the respective urns that a ball to be drawn will be white. Let

$$p = \frac{p_1 + p_2 + \cdots + p_s}{s}. \tag{1}$$

From (1) it follows that the mathematical expectation sp of white balls in a set of s obtained one from each urn is exactly equal to the mathematical expectation of white balls in drawing s balls with a constant probability p of success. The standard deviation σ_P of the theoretical distribution of the number of white balls per set of s is related to the standard deviation $\sigma_B = (spq)^{1/2}$ of a hypothetical Bernoulli distribution with a constant probability p of success, by the equation

$$\sigma_P^2 = spq - \sum_{x=1}^{x=s} (p_x - p)^2 = \sigma_B^2 - \sum_{x=1}^{x=s} (p_x - p)^2, \tag{2}$$

where p is equal to the mean value of $p_1, p_2, \ldots, p_s$. To prove this we start with (1) and recall that sp is the arithmetic mean of the number of white balls in any set of s under the theoretical distribution.

Let us consider next the standard deviation σ of white balls in the theoretical series of s balls. The square of the standard deviation of the frequency of white balls in drawing a single ball with the chance p_t that it will be white is given by $\sigma_t^2 = p_t q_t$, that is, by making $s=1$ in $sp_t q_t$.

When the probabilities $p_1, p_2, \ldots, p_s$ are independent of one another, it follows from (16), page 137, that

$$\sigma^2 = \sigma_1^2 + \sigma_2^2 + \cdots + \sigma_s^2 ,$$

where $\sigma_1, \sigma_2, \ldots, \sigma_s$ are the standard deviations of white balls in drawing one ball from each urn corresponding to probabilities $p_1, p_2, \ldots, p_s$, respectively, and σ is the standard deviation of white balls among the s balls together drawn one from each urn.

Hence, we have

$$(3) \qquad \sigma^2 = p_1 q_1 + p_2 q_2 + \cdots + p_s q_s = \sum_{t=1}^{t=s} p_t q_t .$$

But

$$p_t = p + (p_t - p) , \qquad q_t = q - (p_t - p) .$$

Hence

$$p_t q_t = pq - (p_t - p)(p - q) - (p_t - p)^2$$

and

$$(4) \quad \sum_{t=1}^{t=s} p_t q_t = spq - \sum_{t=1}^{t=s} (p_t - p)^2 , \text{ since } \sum_{t=1}^{t=s} (p_t - p) = 0 .$$

Hence, we have established (2), from which it follows at once that the standard deviation of a Poisson series is less than that of the corresponding Bernoulli series with constant probability of success equal to the arithmetic mean of the variable probabilities of success.

To give an illustration of a Poisson series, conceive of n populated districts. Each district is to consist of s subdivisions for which the probability of death at a given age varies from one subdivision to another, but in which the series of s probabilities are identical from district to district. To illustrate further this type of distribution, construct an urn schema consisting of 10 urns each of which contains 15 balls, and in which the number of white balls in the respective urns is 3, 4, 5, 6, 7, 8, 9, 10, 11, 12. The arithmetic mean of the probabilities of drawing a white ball is 1/2. A set of 10 is obtained by drawing one ball from each urn. Then each ball is returned to the urn from which it was drawn, and a second set of 10 is drawn. This process is continued until we have 1,000 sets of 10. The resulting frequency distribution of the number of white balls is a Poison distribution.

53. **Lexis series.** To give a statistical illustration of a Lexis series, conceive of n populated districts in each of which the probability of death is constant for men of given age, but is variable from district to district.

To develop the theory of the Lexis distribution we draw s balls one at a time from an urn U_1 with a constant probability p_1 of getting a white ball, from U_2 with a constant probability $p_2, \ldots,$ from U_n with a constant probability p_n.

The mathematical expectation of white balls in thus drawing ns balls is $sp_1+sp_2+\cdots+sp_n=nsp$, where $p=(1/n)(p_1+p_2+\cdots+p_n)$ is the arithmetic mean of the probabilities $p_1, p_2, \ldots, p_n$.

Since nsp is the mathematical expectation of white balls in samples of ns balls, the mathematical expectation in samples of s balls one at a time from a random urn is sp. This value sp is identical to the mathematical expectation of white balls in samples of s balls of a Bernoulli series with a constant probability p.

Since p_t is the probability that a ball to be drawn from urn U_t will be white, the expected value of the square of the standard deviation of the number of white balls in samples of s drawn from U_t is sp_tq_t. In other words, sp_tq_t is the mean square of the deviations of white balls from sp_t in samples of s drawn from U_t. If the deviations were measured from sp instead of sp_t, it follows from the theorem (p. 21) for changing the origin or axis of second moments that the mean square of the deviations would be

$$(5) \qquad sp_tq_t+(sp_t-sp)^2 \,.$$

Suppose this mean value of the squares of deviations were obtained from N samples of s each. Then

$$(6) \qquad Nsp_tq_t+Ns^2(p_t-p)^2$$

would be the expected value of the sum of squares of the deviations from sp in the N samples of s drawn from U_t.

By adding together the expression (6) for $t=1, 2, \ldots, n$, we have

$$(7) \qquad Ns\sum_{t=1}^{t=n} p_tq_t+Ns^2\sum_{t=1}^{t=n}(p_t-p)^2$$

for the expected value of the sum of squares of the deviations from sp for the n urns. In obtaining (7), we have drawn in all Nn sets of s balls of which N sets are from each urn.

The mean-square deviation from sp of the number of white balls in samples of s thus taken from the n urns $U_1, U_2, \ldots, U_n$ is then obtained by dividing (7) by the number of sets Nn. This gives

$$\sigma_L^2 = \frac{s}{n}\sum_{t=1}^{t=n} p_t q_t + \frac{s^2}{n}\sum_{t=1}^{t=n}(p_t - p)^2 .$$

From (4) above

$$\sum_{t=1}^{t=n} p_t q_t = npq - \sum_{t=1}^{t=n}(p_t - p)^2 ,$$

and hence

$$\text{(8)} \qquad \sigma_L^2 = spq + \frac{s^2 - s}{n}\sum_{t=1}^{t=n}(p_t - p)^2 = \sigma_B^2 + \frac{s^2 - s}{n}\sum_{t=1}^{t=n}(p_t - p)^2 .$$

It should be observed from (8) that the standard deviation of a Lexis distribution is greater than that of a Bernoulli distribution based on a constant probability p which is equal to the mean value of the given probabilities $p_1, p_2, \ldots, p_n$.

54. **The Lexis ratio.** Let σ' be the standard deviation of a series of relative frequencies obtained by experiment from statistical data. On the hypothesis of a Bernoulli distribution the theoretical value of the standard deviation is $\sigma_B' = (pq/s)^{1/2}$ where p is the probability of success in any single trial. The ratio

$$L = \frac{\sigma'}{\sigma_B'} = \frac{\sigma}{\sigma_B}$$

is called the Lexis ratio, where $\sigma = s\sigma'$ and $\sigma_B = s\sigma'_B$. When $L = 1$,[66] the series of relative frequencies is said to have *normal* dispersion. When $L < 1$, the series is said to have *subnormal* dispersion. When $L > 1$, the series is said to have *supernormal* dispersion. Illustrative applications of the Lexis ratio to statistical data are readily available.[67]

From the nature of the Lexis theory it is fairly obvious, as implied in the introduction to this chapter, that

TABLE I

State	Births[38]	Deaths per 1,000
California	65,457	66
Connecticut	33,471	72
Indiana	66,544	70
Kansas	40,477	61
Kentucky	62,941	58
Minnesota	57,185	58
North Carolina	61,348	66
Virginia	48,535	68
Wisconsin	61,352	72
Arithmetic mean	55,257	65.7

the application of the theory to particular statistical data involves breaking up the aggregate into a number of subsets according to some appropriate scheme of classification which would ordinarily depend on much knowledge of the material which is the subject of the investigation. Then we are concerned not only with a frequency ratio for the entire aggregate, but also with the stability of frequency ratios among the subsets. The dispersion of frequency ratios is calculated and compared with the expected value in the case of a Bernoulli distribution.

As an example, let us consider the dispersion of death-rates of white infants under one year of age in registration states[68] of the United States in which the number of births per year of white children is between 33,000 and 67,000 (see Table I). This restriction is placed on the selection of states so that the number of instances per set has only a moderate amount of variability.

In most of the practical problems of statistics the exact values of the underlying probabilities are unknown and the best substitutes available are the approximate values of the probabilities given by available relative frequencies. Substituting these frequency ratios as approximations for p and q, we find the Bernoulli standard deviation from the formula $\sigma'_B = (pq/s)^{1/2}$. We then compare σ'_B with the standard deviation obtained directly from the data. The simple arithmetic mean of the death-rates is 65.7 per 1,000, and their standard deviation (without weighting) is 5.21 per 1,000. If these infantile death-rates constituted a Bernoulli distribution with a number of instances equal to the average number of births, 55,257 in each case, we should have

$$\sigma'_B = \left(\frac{pq}{s}\right)^{1/2} = \left[\frac{(.0657)(.9343)}{55,257}\right]^{1/2}$$

$$= .00105 \text{ per person}, \qquad = 1.05 \text{ per } 1,000 .$$

Hence, the Lexis ratio is

$$L = \frac{5.21}{1.05} = 4.96 .$$

Hence the dispersion is supernormal, and we have strong support for the inference that there is a significant variation in infant mortality from one of these states to another. The full interpretation of this fact would require much knowledge of the sources of the material.

A reasonable plan for the determination of the maximum district over which the infantile death-rates are essentially constant seems to involve breaking the aggregate of instances into subsets in a variety of ways and then testing results as above. Some measure of doubt will remain, but this procedure encourages the kind of analysis that gives strong support to induction.

CHAPTER VII

A DEVELOPMENT OF THE GRAM-CHARLIER SERIES

55. Introduction. In § 56 we shall attempt to show (cf. p. 65) that a certain line of development[24] of the binomial distribution suggests the use of the Gram-Charlier Type A series as a natural extension of the De Moivre-Laplace approximation and the Type B series as a natural extension of the Poisson exponential approximation considered in Chapter II. Then in §§ 57–58 we shall develop methods for the determination of the parameters in terms of moments of the observed frequency distribution, thus deriving certain results stated without proof in § 19 and § 21.

56. On a development of Type A and Type B from the law of repeated trials. As in the De Moivre-Laplace theory, we consider the probability that in a sample of s individuals, taken at random from an unlimited supply, r individuals will have a certain attribute. That is, the probability we wish to represent is given by

$$B(r)=\frac{s!}{r!\,(s-r)!}\,p^r q^{s-r}\,,$$

and we shall use a function of the form

$$B_0(x)=\frac{1}{2\pi}\int_{-\pi}^{\pi}\theta(w)e^{-xwi}\,dw\,, \tag{1}$$

for interpolation between the values $B(r)$, where $i^2=-1$ and

$$\theta(w)=(pe^{wi}+q)^s=\sum_{r=0}^{r=s} B(r)e^{rwi}. \tag{2}$$

In the terminology of Laplace, $\theta(w)$ is the generating function of the sequence $B(r)$.

We shall first show that $B_0(x)=B(m)$ when x is a positive integer m. To prove this, substitute $\theta(w)$ from (2) in (1) and integrate. This gives

$$\begin{aligned} B_0(x) &= \sum_{r=0}^{r=s} B(r)\frac{\sin(r-x)\pi}{(r-x)\pi} \\ &= B(0)\frac{\sin(-x\pi)}{-x\pi}+B(1)\frac{\sin(1-x)\pi}{(1-x)\pi} \\ &\qquad +\cdots+B(s)\frac{\sin(s-x)\pi}{(s-x)\pi}. \end{aligned}$$

When $x=m$ is a positive integer, each term but one of the right member vanishes and this one has the value $B(m)$. Accordingly, $B_0(m)=B(m)$.

Thus formula (1) gives exactly the terms of the expansion of $(p+q)^s$ for positive integral values $x=m$. It may be considered an interpolation formula for values of x between the integral values.

We shall be interested in two developments of this interpolation formula. The first is based on the development of $\log\theta(w)$ in powers of w, and the second on the development in powers of p. The resulting types of development are known as the Type A and Type B series, respectively.

DEVELOPMENT OF TYPE A

From the form of $\theta(w)$ in (2), we have

$$(3) \qquad \frac{d \log \theta(w)}{dw} = \frac{ispe^{wi}}{q+pe^{wi}}.$$

Develop the right-hand member of (3) in powers of w and we obtain

$$\frac{d \log \theta(w)}{dw} = isp[1+qwi-\tfrac{1}{2}q(p-q)(wi)^2+\cdots].$$

Thus we have by integration, remembering that $\theta(0)=1$,

$$\log \theta(w) = s\left[pwi+\frac{1}{2!}pq(wi)^2-\frac{1}{3!}pq(p-q)(wi)^3+\cdots\right],$$

or writing

$$(4) \qquad \theta(w) = e^{b_1wi+\frac{b_2}{2!}(wi)^2+\frac{b_3}{3!}(wi)^3+\cdots},$$

we have

$$(5) \quad b_1 = sp, \qquad b_2 = spq, \qquad b_3 = -spq(p-q), \cdots.$$

We now write

$$(6) \quad \theta(w) = e^{b_1wi-b_2w^2/2}[1-A_3(wi)^3+A_4(wi)^4-\cdots].$$

Since it follows from (2) that $\theta(w)$ is an entire function of w, the series in brackets in the right member of (6) converges since it is the quotient of an entire function $\theta(w)$ by an exponential factor with no singularities in the finite part of the plane.

From (4), (5), and (6) we have

$$(7) \quad A_3 = \tfrac{1}{6} spq(p-q) , \qquad A_4 = \tfrac{1}{24} spq(1-6pq) , \cdots .$$

Inserting $\theta(w)$ from (6) in (1), we have

$$(8) \quad B_0(x) = \frac{1}{2\pi} \int_{-\pi}^{\pi} dw e^{-(x-b_1)wi - b_2 w^2/2} [1 - A_3(wi)^3 + A_4(wi)^4 - \cdots] .$$

If we write

$$(9) \qquad \Omega(x) = \frac{1}{2\pi} \int_{-\pi}^{\pi} dw e^{-(x-b_1)wi - b_2 w^2/2} ,$$

we have from (8),

$$(10) \quad B_0(x) = \Omega(x) + A_3 \frac{d^3\Omega(x)}{dx^3} + A_4 \frac{d^4\Omega(x)}{dx^4} + \cdots .$$

If, however, b_2 is not small, we may use in place of $\Omega(x)$ the function $\phi(x)$ defined as

$$\phi(x) = \frac{1}{2\pi} \int_{-\infty}^{+\infty} dw e^{-(x-b_1)wi - b_2 w^2/2}$$

by changing the limits of integration from $\pm\pi$ to $\pm\infty$. Moreover, we shall prove that

$$(11) \qquad \phi(x) = \frac{1}{(2\pi b_2)^{1/2}} e^{-(x-b_1)^2/2b_2} .$$

To prove this, we write

$$\phi(x) = \frac{1}{2\pi} \int_{-\infty}^{+\infty} e^{-b_2 w^2/2} \cos [w(x-b_1)] dw - \frac{i}{2\pi} \int_{-\infty}^{+\infty} e^{-b_2 w^2/2} \sin [w(x-b_1)] dw .$$

The second term vanishes because the sine is an odd function. Since the cosine is an even function, we may write

$$(12) \qquad \phi(x)=\frac{1}{\pi}\int_0^\infty e^{-b_2w^2/2}\cos\,[w(x-b_1)]dw\,.$$

Differentiation with regard to x gives

$$\frac{d\phi(x)}{dx}=-\frac{1}{\pi}\int_0^\infty e^{-b_2w^2/2}\,w\sin\,[w(x-b_1)]dw\,.$$

Integrate the right-hand member by parts and we have

$$\frac{d\phi(x)}{dx}=-\frac{(x-b_1)}{b_2\pi}\int_0^\infty e^{-b_2w^2/2}\cos\,[w(x-b_1)]dw$$

$$=-\frac{(x-b_1)}{b_2}\phi(x)\,.$$

Then by integration,

$$(13) \qquad \phi(x)=Ae^{-(x-b_1)^2/2b_2}\,.$$

To find A, let $x=b_1$ in (12) and (13). This gives the well known definite integral

$$A=\frac{1}{\pi}\int_0^\infty e^{-b_2w^2/2}\,dw=\frac{1}{(2\pi b_2)^{1/2}}\,.$$

Hence, we have

$$\phi(x)=\frac{1}{(2\pi b_2)^{1/2}}\,e^{-(x-b_1)^2/2b_2}$$

as given in (11).

Therefore we may write in place of (10) the Type A series

$$(14)\quad B_0(x)=\phi(x)+A_3\frac{d^3\phi(x)}{dx^3}+A_4\frac{d^4\phi(x)}{dx^4}+\cdots.$$

where

$$\phi(x)=\frac{1}{\sigma(2\pi)^{1/2}}e^{-(x-b_1)^2/2\sigma^2},$$

if $\sigma^2=b_2$.

To study the degree of approximation secured in changing the limits of integration from $\pm\pi$ to $\pm\infty$ in passing from $\Omega(x)$ to $\phi(x)$, we observe that

$$\phi(x)-\Omega(x)=\frac{1}{\pi}\int_\pi^\infty dwe^{-\sigma^2w^2/2}\cos[(x-b_1)]$$

and hence

$$[\phi(x)-\Omega(x)]<\frac{1}{\pi}\int_\pi^\infty dwe^{-\sigma^2w^2/2}=\frac{1}{\sigma\pi}\int_{\sigma\pi}^\infty e^{-\lambda^2/2}\,d\lambda,$$

if $\lambda^2=\sigma^2w^2$.

Hence, the difference approaches zero very rapidly with increasing values of σ as may be seen by using the values of the last integral written corresponding to values of $\lambda=1, 2, 3, 4, \ldots$, in a table of this probability integral. A similar examination for the derivatives of Ω and ϕ will show that their differences similarly approach zero.

DEVELOPMENT OF TYPE B

To develop (2) in powers of p, we first write, since $p+q=1$,

$$(15)\left\{\begin{aligned}\frac{d\log\theta(w)}{dw}&=\frac{ispe^{wi}}{1-p(1-e^{wi})}\\&=ispe^{wi}[1+p(1-e^{wi})+p^2(1-e^{wi})^2+\cdots],\end{aligned}\right.$$

a convergent series since

$$|p(1-e^{wi})|<1.$$

Since $\theta(0)=1$, we obtain by integration

$$(16)\quad \log\theta(w)= -sp\left[1-e^{wi}+\frac{p}{2}(1-e^{wi})^2+\frac{p^2}{3}(1-e^{wi})^3+\cdots\right].$$

Hence, writing

$$(17)\quad \theta(w)=e^{-sp(1-e^{wi})}[1+B_2(1-e^{wi})^2+B_3(1-e^{wi})^3+\cdots],$$

we have

$$B_2=-\frac{sp^2}{2},\qquad B_3=-\frac{sp^3}{3},\qquad B_4=-\frac{sp^4}{4}+\frac{s^2p^4}{8},\cdots.$$

Now, from (1) and (17),

$$(18)\quad B_0(x)=\frac{1}{2\pi}\int_{-\pi}^{\pi}dwe^{-xwi-sp(1-e^{wi})}[1+B_2(1-e^{wi})^2 +B_3(1-e^{wi})^3+\cdots].$$

Let

$$\psi(x)=\frac{1}{2\pi}\int_{-\pi}^{\pi}dwe^{-xwi-sp(1-e^{wi})}=\int_{-\pi}^{\pi}Q(w,x)dw.$$

Then let

$$\Delta\psi(x)=\psi(x)-\psi(x-1)=\frac{1}{2\pi}\int_{-\pi}^{\pi}dwe^{-xwi-sp(1-e^{wi})} -\frac{1}{2\pi}\int_{-\pi}^{\pi}dwe^{-(x-1)wi-sp(1-e^{wi})}$$

$$=\int_{-\pi}^{\pi}(1-e^{wi})Q(w,x)dw.$$

Then
$$\Delta^2\psi(x) = \int_{-\pi}^{\pi} (1-e^{wi})^2 Q(w, x)\,dw\,,$$
$$\Delta^3\psi(x) = \int_{-\pi}^{\pi} (1-e^{wi})^3 Q(w, x)\,dw\,,$$
$$\cdots\cdots\cdots\cdots\cdots\cdots$$

Hence, we have

$$B_0(x) = \psi(x) + B_2\Delta^2\psi(x) + B_3\Delta^3\psi(x) + \cdots. \tag{19}$$

To give other forms to

$$\psi(x) = \frac{1}{2\pi}\int_{-\pi}^{\pi} e^{-xwi - sp(1-e^{wi})}\,dw\,,$$

we may write

$$\psi(x) = \frac{e^{-sp}}{2\pi}\int_{-\pi}^{\pi} e^{-xwi + spe^{wi}}\,dw$$
$$= \frac{e^{-sp}}{2\pi}\int_{-\pi}^{\pi} e^{-xwi}\left[1 + sp\,e^{wi} + \frac{s^2p^2}{2!}e^{2wi} + \cdots\right] dw$$
$$= \frac{e^{-sp}}{2\pi}\left[\int_{-\pi}^{\pi} e^{-xwi}\,dw + sp\int_{-\pi}^{\pi} e^{-wi(x-1)}dw + \frac{s^2p^2}{2!}\int_{-\pi}^{\pi} e^{-wi(x-2)}\,dw + \cdots\right]$$
$$= \frac{e^{-sp}}{\pi}\left[\frac{\sin x\pi}{x} + sp\,\frac{\sin(x-1)\pi}{x-1} + \cdots + \frac{(sp)^r}{r!}\,\frac{\sin(x-r)\pi}{x-r} + \cdots\right] \tag{20}$$
$$= \frac{e^{-sp}}{\pi}\sin\pi x\left[\frac{1}{x} - \frac{sp}{x-1} + \frac{s^2p^2}{2!\,(x-2)} - \cdots + (-1)^r\,\frac{(sp)^r}{r!}\,\frac{1}{x-r} + \cdots\right]$$

$$(21) \quad = e^{-\lambda} \frac{\sin \pi x}{\pi} \left[\frac{1}{x} - \frac{\lambda}{x-1} + \frac{\lambda^2}{(x-2)2!} - \cdots \right.$$

$$\left. + \frac{(-1)^r \lambda^r}{(x-r)r!} + \cdots \right],$$

if sp is replaced by λ.

The foregoing analytical processes can be easily justified by the use of the properties of uniformly convergent series. When x approaches an integer r, it is easily seen from (20) that each term approaches zero except the term

$$\frac{e^{-sp}}{\pi} \frac{(sp)^r}{r!} \frac{\sin (x-r)\pi}{x-r},$$

and this term has as its limit the Poisson exponential

$$\frac{e^{-sp}(sp)^r}{r!} = \frac{e^{-\lambda}\lambda^r}{r!}.$$

The formula (21) may therefore be regarded as defining the Poisson exponential $e^{-\lambda}\lambda^x/x!$ for non-integral values of x.

The development in series (19) is useful only when p is so small that sp is not large, say $sp \leqq 10$, s being a large number. In this case, sp is likely to be too small to allow an expansion in a Type A series. Otherwise, the development in Type A is better suited to represent the terms of the binomial series.

While the above demonstration is limited to the representation of the law of probability given by terms of a binomial, Wicksell has gone much further in the paper cited above in showing a line of development which suggests the use of the Gram-Charlier series for the representation of the law of probability given by terms of the hypergeometric series, thus representing the law of probability which gives the basis of the Pearson system of generalized frequency curves. Unfortunately, the demonstration of this extension would require somewhat more space devoted to formal analysis than seems desirable in the present monograph. Hence we merely state the above fact without a demonstration.

57. The values of the coefficients of the Type A series obtained from the biorthogonal property. If in (14) we measure x from the centroid as an origin and in units equal to the standard deviation, σ, we may write in place of (14)

$$F(x)=\phi(x)+a_3\phi^{(3)}(x)+a_4\phi^{(4)}(x)+\cdots+a_n\phi^{(n)}(x)+\cdots, \tag{22}$$

where

$$\phi(x)=\frac{1}{\sigma(2\pi)^{1/2}}e^{-x^2/2},$$

and $\phi^{(n)}(x)$ is the nth derivative of $\phi(x)$ with respect to x.

The coefficients $a_n(n=0, 3, 4, \ldots .)$ in the Type A series may be easily expressed in terms of moments of area under the given frequency curve about the centroidal ordinate because the functions $\phi^{(n)}(x)$ and the Hermite polynomials $H_m(x)$ defined by the equation

$$\phi^{(m)}(x)=(-1)^m H_m(x)\phi(x)$$

form a biorthogonal system. Thus,

$$(23) \qquad \int_{-\infty}^{\infty} \phi^{(n)}(x) H_m(x) dx = 0 \qquad (m \neq n),$$

$$(24) \qquad \text{and} \quad \int_{-\infty}^{\infty} \phi_{(x)}^{(n)} H_m(x) dx = \frac{(-1)^n n!}{\sigma} \qquad (m = n),$$

and this biorthogonal property affords a simple method of determining the coefficients in the Type A series.

To prove (23) and (24) we may write

$$(25) \quad \begin{cases} \int_{-\infty}^{\infty} \phi^{(n)}(x) H_m(x) dx = (-1)^n \int_{-\infty}^{\infty} \phi(x) H_n(x) H_m(x) dx \\ \qquad = (-1)^{m+n} \int_{-\infty}^{\infty} \phi^{(m)}(x) H_n(x) dx . \end{cases}$$

Integration by parts gives

$$\int_{-\infty}^{\infty} \phi^{(n)}(x) H_m(x) dx = \left[\phi^{(n-1)}(x) H_m(x)\right]_{-\infty}^{\infty}$$

$$- \int_{-\infty}^{\infty} \phi^{(n-1)}(x) H_m'(x) dx = - \int_{-\infty}^{\infty} \phi^{(n-1)}(x) H_m'(x) dx .$$

Continuing until we have performed $m+1$ successive integrations by parts, we obtain, assuming $n > m$,

$$\int_{-\infty}^{\infty} \phi^{(n)}(x) H_m(x) dx =$$

$$(-1)^{m+1} \int_{-\infty}^{\infty} \phi^{(n-m-1)}(x) H_m{}^{(m+1)}(x) dx ,$$

where $H_m^{(m+1)}(x)$ is the $(m+1)$th derivative of $H_m(x)$. Since $H_m(x)$ is a polynomial of degree m in x, its $(m+1)$th derivative vanishes and we have

$$\int_{-\infty}^{\infty} \phi^{(n)}(x)H_m(x)dx=0 \tag{26}$$

for $n>m$. But from the form of (25), it is obvious that we could equally well prove (26) for $m>n$. For $m=n$, we proceed as above with m successive integrations. We then have, if we replace m by n,

$$\begin{aligned}\int_{-\infty}^{\infty} \phi^{(n)}(x)H_n(x)dx &= (-1)^n\int_{-\infty}^{\infty} \phi^{(n-n)}(x)H_n^{(n)}(x)dx\\ &= (-1)^n\int_{-\infty}^{\infty} \phi(x)H_n^{(n)}(x)dx.\end{aligned}$$

But the nth derivative $H_n^{(n)}(x)$ of the polynomial $H_n(x)$ is equal to $n!$. Hence,

$$\left\{\begin{aligned}\int_{-\infty}^{\infty} \phi^{(n)}(x)H_n(x)dx &= (-1)^n n!\int_{-\infty}^{\infty} \phi(x)dx\\ &= \frac{(-1)^n n!}{\sigma(2\pi)^{1/2}}\int_{-\infty}^{\infty} e^{-x^2/2}dx=\frac{(-1)^n n!}{\sigma}\end{aligned}\right. \tag{27}$$

By multiplying both members of (22) by $H_n(x)$ and integrating under the assumption that the series is uniformly convergent, we have

$$\int_{-\infty}^{\infty} F(x)H_n(x)dx=a_n\int_{-\infty}^{\infty} \phi^{(n)}(x)H_n(x)dx=(-1)^n\frac{a_n n!}{\sigma}$$

since by application of (26) all terms of the right-hand member vanish except the one with the coefficient a_n. Hence,

$$a_n = \frac{\sigma(-1)^n \int_{-\infty}^{\infty} F(x)\, H_n(x) dx}{n!} . \tag{28}$$

Moreover, to determine a_n numerically for an observed frequency distribution we replace $F(x)$ in (28) by the observed frequency function $f(x)$.

For purposes of numerical application, let us now change back from the standard deviation as a unit to measuring x in the ordinary unit of measurement (feet, pounds, etc.) involved in the problem, but still keep the origin at the centroid. This means that we replace x in (28) by x/σ. If in these units $f(x)$ gives the observed frequency distribution, we may write in place of (28)

$$\begin{cases} a_n = \sigma \dfrac{(-1)^n}{n!} \displaystyle\int_{-\infty}^{\infty} f(x)\, H_n(x/\sigma)\, dx/\sigma \\ \quad = \dfrac{(-1)^n}{n!} \displaystyle\int_{-\infty}^{\infty} f(x)\, H_n(x/\sigma)\, dx . \end{cases} \tag{29}$$

Since $H_n(x/\sigma)$ is a polynomial of degree n in x, the coefficients a_n are thus given in terms of moments of area under the observed frequency curve. It is then fairly obvious that the determination of the moments of area under the frequency curve plays an important part in the Gram-Charlier system as well as in the Pearson system.

58. **The values of the coefficients of Type A series obtained from a least-squares criterion.** It may be proved by following J. P. Gram that the value of any coefficient a_n obtained in § 57 by the use of the biorthogonal property

is the same as that obtained by finding the best approximation to $f(x)$, in the sense of a certain least-squares criterion, by the first m terms of the series ($m \geqq n$). To prove this statement, we may proceed as follows: Consider the series

$$F(x) = a_0\phi(x) + a_1\phi^{(1)}(x) + \cdots + a_m\phi^{(m)}(x)$$

for the representation of an observed frequency function $f(x)$. The least-squares criterion[25] that

$$V = \int_{-\infty}^{\infty} \frac{1}{\phi(x)} [f(x) - F(x)]^2 \, dx \tag{30}$$

shall be a minimum leads to values of coefficients given in § 19.

To prove this, we square the binomial $f(x) - F(x)$ and differentiate partially with regard to the parameters $a_0, a_3, \ldots, a_m$. This gives

$$\frac{\partial V}{\partial a_n} = -2\frac{\partial}{\partial a_n}\int_{-\infty}^{\infty} \frac{f(x)F(x)dx}{\phi(x)} + \frac{\partial}{\partial a_n}\int_{-\infty}^{\infty} [F(x)]^2 \frac{1}{\phi(x)} dx$$

$$= 2(-1)^{n+1}\int_{-\infty}^{\infty} f(x)\, H_n(x)dx + 2a_n \int_{-\infty}^{\infty} [H_n(x)]^2\, \phi(x)dx$$

since

$$\int_{-\infty}^{\infty} \frac{1}{\phi(x)} [F(x)]^2 dx = \int_{-\infty}^{\infty} \frac{1}{\phi(x)} [\sum a_n \phi^{(n)}(x)]^2 \, dx$$

$$= \int_{-\infty}^{\infty} \{a_0^2[H_0(x)]^2 + a_1^2[H_1(x)]^2 + \cdots + a_m^2[H_m(x)]^2\}\phi(x)dx \,,$$

the product terms vanishing because of (26).

Making $\partial V/\partial a_n = 0$, we have

$$(31)\ 2(-1)^{n+1}\int_{-\infty}^{\infty} f(x)\, H_n(x)dx + 2a_n \int_{-\infty}^{\infty} [H_n(x)]^2\, \phi(x)dx = 0\,.$$

But

$$(32)\quad \int_{-\infty}^{\infty} [H_n(x)]^2 \phi(x)dx = (-1)^n \int_{-\infty}^{\infty} \phi^{(n)}(x) H_n(x)dx = \frac{n!}{\sigma}\cdot$$

From (31) and (32), we have

$$a_n = \frac{\sigma(-1)^n}{n!} \int_{-\infty}^{\infty} f(x)\, H_n(x)dx\,,$$

which is identical with the value obtained by the use of the biorthogonal property.

59. **The coefficients of a Type B series.** In considering the determination of the coefficients $c_0, c_1, c_2, \ldots$, of the Type B series, we shall restrict our treatment to a distribution of equally distant ordinates at non-negative integral values of x, and shall for simplicity consider the representation by the first three terms of the series. That is, we write

$$F(x) = c_0\psi(x) + c_1\Delta\psi(x) + c_2\Delta^2\psi(x)\,,$$

where

$$\psi(x) = \frac{e^{-\lambda}\lambda^x}{x!}$$

for $x = 0, 1, 2, \ldots$. Let $f(x)$ give the ordinates of the observed distribution of relative frequencies, so that $\sum f(x) = 1$.

Equating sums of ordinates and first and second moments μ_1' and μ_2' of ordinates of the theoretical and observed distributions, we may now determine the coefficients approximately from the equations:

$$(33)\begin{cases}\sum\ [c_0\psi(x)+c_1\Delta\psi(x)+c_2\Delta^2\psi(x)]=\sum f(x)=1\\ \sum x\ [c_0\psi(x)+c_1\Delta\psi(x)+c_2\Delta^2\psi(x)]=\sum xf(x)=\mu_1'\\ \sum x^2[c_0\psi(x)+c_1\Delta\psi(x)+c_2\Delta^2\psi(x)]=\sum x^2f(x)=\mu_2'\end{cases}$$

Before solving these equations for c_0, c_1, and c_2, we may simplify by the substitution of certain values which are close approximations when we are dealing with large numbers. Thus, we recall that we have derived in § 14, Chapter II, the following approximations:

$$\sum\psi(x)=1\ ,$$

$$\sum mP_m=\sum x\psi(x)=\lambda\ ,$$

$$\mu_2'=\sum x^2\psi(x)=\lambda+\lambda^2\ .$$

We may next easily obtain the following further approximate values:

$$\sum\Delta\psi(x)\ =\sum[\psi(x)-\psi(x-1)]=1-1=0\ ,$$

$$\sum\Delta^2\psi(x)=\sum[\psi(x)-2\psi(x-1)+\psi(x-2)]=0\ ,$$

$$\begin{aligned}\sum x\Delta\psi(x)&=\sum x[\psi(x)-\psi(x-1)]\\ &=\sum x\psi(x)-\sum(x-1)\psi(x-1)-\psi(x-1)\ ,\\ &=\lambda-\lambda-1=-1\ .\end{aligned}$$

Similarly, it is easily shown that

$$\sum x\Delta^2\ \psi(x)=0,$$

$$\sum x^2\Delta\ \psi(x)=-2\lambda-1\ ,$$

and

$$\sum x^2 \Delta^2 \psi(x) = 2\ .$$

Substituting these values in equations (33), we obtain

$$c_0 = 1\ , \qquad \lambda c_0 - c_1 = \mu_1'\ ,$$
$$(\lambda + \lambda^2) c_0 - (2\lambda + 1) c_1 + 2c_2 = \mu_2'\ .$$

If we take $\lambda = \mu_1'$, we have the coefficient $c_1 = 0$.

Then expressing the second moment μ_2' in terms of the second moment μ_2 about the mean by the relation

$$\mu_2' = \mu_2 + \lambda^2\ ,$$

we have

$$\lambda + \lambda^2 + 2c_2 = \mu_2 + \lambda^2\ , \qquad c_2 = \tfrac{1}{2}(\mu_2 - \lambda)\ .$$

Hence, we write

$$F(x) = \psi(x) + \tfrac{1}{2}(\mu_2 - \lambda)\Delta^2 \psi(x)\ ,$$

when λ is taken equal to the first moment μ_1', which is the arithmetic mean of the values of the given variates.

It is fairly obvious that this application of moments to finding values of the coefficients can be extended to more terms if they were needed in dealing with actual data.

NOTES

1. Page 1. Émile Borel, *Éléments de la théorie des probabilités*, p. 167; *Le hasard*, p. 154.

2. Page 23. Julian L. Coolidge, *An introduction to mathematical probability* (1925), pp. 13–32.

3. Page 30. Tchebycheff, *Des valeurs moyennes*. Journal de Mathématique (2), Vol. 12 (1867), pp. 177–84.

4. Page 30. M. Bienaymé, *Considérations à l'appui de la découverte de Laplace sur la loi de probabilité dans la méthode des moindres carrés*, Comptes Rendus, Vol. 37 (1853), pp. 309–24.

5. Page 31. E. L. Dodd, *The greatest and the least variate under general laws of error*, Transactions of the American Mathematical Society, Vol. 25 (1923), pp. 525–39.

6. Pages 31 and 47. Some writers call this theorem the "Bernoulli theorem" and others call it the "Laplace theorem." It has been shown recently by Karl Pearson that most of the credit for the theorem should go to De Moivre rather than to Bernoulli. For this reason we call the theorem the "De Moivre-Laplace theorem" rather than the "Bernoulli-Laplace theorem." See *Historical note on the origin of the normal curve of errors*, Biometrika, Vol. 16 (1924), p. 402; also James Bernoulli's theorem, Biometrika, Vol. 17 (1925), p. 201.

7. Page 32. For a proof see Coolidge, *An introduction to mathematical probability*, pp. 38–42.

8. Pages 37 and 68. James W. Glover, *Tables of applied mathematics* (1923), pp. 392–411.

9. Page 37. Karl Pearson, *Tables for statisticians and biometricians* (1914), pp. 2–9.

10. Page 39. Poisson, *Recherches sur la probabilité des jugements*, Paris, 1837, pp. 205 ff.

11. Page 39. Bortkiewicz, *Das Gesetz der kleinen Zahlen*, Leipzig, 1898.

12. Page 48. For various proofs of the normal law, see David Brunt *The combination of observations* (1917), pp. 11–24; also Czuber *Beobachtungsfehler* (1891), pp. 48–110.

13. Page 50. Karl Pearson, *Mathematical contributions to the theory of evolution*, Philosophical Transactions, A, Vol. 186 (1895), pp. 343–414.

14. Page 50. Karl Pearson, *Supplement to a memoir on skew variation* Philosophical Transactions, A, Vol. 197 (1901), pp. 443–56.

15. Page 50. Karl Pearson, *Second supplement to a memoir on skew variation*, Philosophical Transactions, A, Vol. 216 (1916), pp. 429–57.

16. Page 60. J. P. Gram, *Om Raekkeudviklinger* (1879) (Doctor's dissertation), Copenhagen, 1879; also *Über die Entwickelung reeller Functionen in Reihen mittelst die Methode der kleinsten Quadrate*, Journal für Mathematik, Vol. 94 (1883), pp. 41–73.

17. Page 60. T. N. Thiele, *Almindelig Iagttagelseslaere*, Copenhagen, 1889; cf. Thiele, *Theory of observations*, 1903.

18. Page 60. F. Y. Edgeworth, *The asymetrical probability-curve*, Philosophical Magazine, Vol. 41 (1896), pp. 90–99; also *The law of error*, Cambridge Philosophical Transactions, Vol. 20 (1904), pp. 36–65, 113–41.

19. Page 60. G. T. Fechner, *Kollektivmasslehre* (ed., G. R. Lipps), 1897.

20. Page 60. H. Bruns, *Über die Darstellung von Fehlergesetzen*, Astronomische Nachrichten, Vol. 143, No. 3429 (1897); also *Wahrscheinlichkeitsrechnung und Kollektivmasslehre*, 1906.

21. Page 60. C. V. L. Charlier, *Über das Fehlergesetz*, Arkiv för Matematik, Astronomi och Fysik, Vol. 2, No. 8 (1905), pp. 1-9; also *Über die Darstellung willkürlicher Funktionen*, Arkiv för Matematik Astronomi och Fysik, Vol. 2, No. 20 (1905), pp. 1–35.

22. Page 60. V. Romanovsky, *Generalization of some types of frequency curves of Professor Pearson*, Biometrika, Vol. 16 (1924), pp. 106–17.

23. Page 63. Wera Myller-Lebedeff, *Die Theorie der Integralgleichungen in Anwendung auf einige Reihenentwicklungen*, Mathematische Annalen, Vol. 64 (1907), pp. 388–416.

24. Pages 65 and 156. S. D. Wicksell, *Contributions to the analytical theory of sampling*, Arkiv för Matematik, Astronomi och Fysik, Vol. 17, No. 19 (1923), pp. 1–46.

25. Pages 67 and 169. In the use of the least-squares criterion that V in (16), §20, and in (30), §58, shall be a minimum, a question naturally arises as to the propriety of weighting squares of deviations with the reciprocal $1/\phi(x)=(2\pi)^{1/2}e^{x^2/2}$ of the normal function. Gram used this weighting without commenting on its propriety so far as the writer has been able to learn. One fairly obvious point in support of the weighting is its algebraic convenience.

26. Page 68. N. R. Jörgensen, *Undersögelser over Frequensflader og Korrelation* (1916), pp. 178–93.

27. Page 74. H. L. Rietz, *Frequency distributions obtained by certain transformations of normally distributed variates*, Annals of Mathematics, Vol. 23 (1922), pp. 292–300.

28. Page 74. S. D. Wicksell, *On the genetic theory of frequency*, Arkiv för Matematik, Astronomi och Fysik, Vol. 12, No. 20 (1917), pp. 1–56.

29. Page 75. E. L. Dodd, *The frequency law of a function of variables with given frequency laws*, Annals of Mathematics, Ser. 2, Vol. 27, No. 1 (1925), pp. 12–20.

30. Page 75. S. Bernstein, *Sur les courbes de distribution des probabilités*, Mathematische Zeitschrift, Vol. 24 (1925), pp. 199–211.

31. Page 81. Francis Galton, Proceedings of the Royal Society, Vol. 40 (1886), Appendix by J. D. Hamilton Dickson, p. 63.

32. Page 81. Karl Pearson, *Mathematical contributions to the theory of evolution III*, Philosophical Transactions, A, Vol. 187 (1896), pp. 253–318.

33. Page 81. G. Udny Yule, *On the significance of the Bravais's formulae for regression, etc.*, Proceedings of the Royal Society, Vol. 60 (1897), pp. 477–89.

34. Page 84. E. V. Huntington, *Mathematics and statistics*, American Mathematical Monthly, Vol. 26 (1919), p. 424.

35. Page 86. H. L. Rietz, *On functional relations for which the coefficient of correlation is zero*, Quarterly Publications of the American Statistical Association, Vol. 16 (1919), pp. 472–76.

36. Page 91. Karl Pearson, *On a correction to be made to the correlation ratio*, Biometrika, Vol. 8 (1911–12), pp. 254–56; see also Student, *The correction to be made to the correlation ratio for grouping*, Biometrika, Vol. 9 (1913), pp. 316–20.

37. Page 92. Karl Pearson, *On a general method of determining the successive terms in a skew regression line*, Biometrika, Vol. 13 (1920–21) pp. 296–300.

38. Page 100. Maxime Bôcher, *Introduction to higher algebra* (1912). p. 33.

39. Page 101. L. Isserlis, *On the partial correlation ratio*, Biometrika Vol. 10 (1914–15), pp. 391–411.

40. Page 101. Karl Pearson, *On the partial correlation ratio*, Proceedings of the Royal Society, A, Vol. 91 (1914–15), pp. 492–98.

41. Page 102. H. L. Rietz, *Urn schemata as a basis for the development of correlation theory*, Annals of Mathematics, Vol. 21 (1920), pp. 306–22.

42. Page 103. A. A. Tschuprow, *Grundbegriffe und Grundprobleme der Korrelationstheorie* (1925).

43. Page 109. H. L. Rietz, *On the theory of correlation with special reference to certain significant loci on the plane of distribution*. Annals of Mathematics (Second Series), Vol. 13 (1912), pp. 195–96.

44. Pages 111 and 112. Karl Pearson, *On the theory of contingency*

and its relation to association and normal correlation, Drapers' Company Research Memoirs (Biometric Series I), (1904), p. 10.

45. Page 111. E. Czuber, *Theorie der Beobachtungsfehler* (1891), pp. 355–82.

46. Page 111. James McMahon, *Hyperspherical goniometry; and its application to correlation theory for N variables*, Biometrika, Vol. 15 (1923), pp. 192–208; paper edited by F. W. Owens after the death of Professor McMahon.

47. Page 112. Karl Pearson, *On the correlation of characters not quantitatively measurable*, Philosophical Transactions, A, Vol. 195 (1900), pp. 1–47.

48. Page 112. Karl Pearson, *On further methods of determining correlation*, Drapers' Company Research Memoirs (Biometric Series IV) (1907), pp. 10–18.

49. Page 112. Warren M. Persons, *Correlation of time series* (Handbook of Mathematical Statistics [1924], pp. 150–65).

50. Page 112. C. Gini, *Nuovi contributi alla teoria delle relazioni statistiche*, Atti del R. Istituto Veneto di S.L.A., Tome 74, P. II (1914–15).

51. Page 112. Louis Bachelier, *Calcul des probabilités* (1912), chaps. 17 and 18.

52. Page 113. Seimatsu Narumi, *On the general forms of bivariate frequency distributions which are mathematically possible when regression and variation are subjected to limiting conditions*, Biometrika, Vol. 15 (1923), pp. 77–88, 209–21.

53. Page 113. Karl Pearson, *Notes on skew frequency surfaces*, Biometrika, Vol. 15 (1923), pp. 222–44.

54. Page 113. Burton H. Camp, *Mutually consistent multiple regression surfaces*, Biometrika, Vol. 17 (1925), pp. 443–58.

55. Page 126. G. Bohlmann, *Formulierung und begründung zweier Hilfssätze der mathematischen Statistik*, Mathematische Annalen, Vol. 74 (1913), pp. 341–409.

56. Page 134. Burton H. Camp, *Problems in sampling*, Journal of the American Statistical Association, Vol. 18 (1923), pp. 964–77.

57. Page 138. Student, *The probable error of a mean*, Biometrika, Vol. 6 (1908–9), pp. 1–25.

58. Page 138. Karl Pearson, *On the distribution of the standard deviations of small samples: Appendix I. To papers by "Student" and R. A. Fisher*, Biometrika, Vol. 10 (1914–15), pp. 522–29.

59. Page 139. R. A. Fisher, *Frequency distribution of the values of the correlation coefficient in samples from an indefinitely large population*, Biometrika, Vol. 10 (1914–15), pp. 507–21.

60. Page 139. H. E. Soper and Others, *On the distribution of the correlation coefficient in small samples. Appendix II to the papers of "Student" and R. A. Fisher*, Biometrika, Vol. 11 (1915–17), pp. 328–413.

61. Page 142. Karl Pearson, *On generalised Tchebycheff theorems in the mathematical theory of statistics*, Biometrika, Vol. 12 (1918–19), pp. 284–96.

62. Page 143. M. Alf. Guldberg, *Sur le théorème de M. Tchebycheff*, Comptes Rendus, Vol. 175 (1922), p. 418; also *Sur quelques inégalités dans le calcul des probabilités*, Vol. 175 (1922), p. 1382. M. Birger Meidell, *Sur un problème du calcul des probabilités et les statistiques mathématiques*, Comptes Rendus, Vol. 175 (1922), p. 806; also *Sur la probabilité des erreurs*, Comptes Rendus, Vol. 176 (1923), p. 280. B. H. Camp, *A new generalization of Tchebycheff's statistical inequality*, Bulletin of the American Mathematical Society, Vol. 28 (1922), pp. 427–32. Seimatsu Narumi, *On further inequalities with possible applications to problems in the theory of probability*, Biometrika, Vol. 15 (1923), p. 245.

63. Page 147. W. Lexis, *Über die Theorie der Stabilität statistischer Reihen*, Jahrbuch für National Ök. u. Statistik, Vol. 32 (1879), pp. 60–98. *Abhandlungen zur theorie der bevölkerungs und moral-statistik*, Kap. V–IX (1903).

64. Page 147. C. V. L. Charlier, *Vorlesungen über die Grundzüge der mathematischen Statistik* (1920), p. 5.

65. Page 147. J. M. Keynes, *A treatise on probability* (1921), p. 393.

66. Page 153. In this connection, the expression "$L \lesseqqgtr 1$" means "$L \lesseqqgtr 1$ apart from chance fluctuations."

67. Page 153. *Handbook of Mathematical Statistics* (1924), pp. 88–91 C. V. L. Charlier, *Vorlesungen über die Grundzüge der mathematischen Statistik* (1920), pp. 38–42.

68. Page 154. *Birth statistics for the registration area of the United States* (1921), p. 37.

INDEX

(Numbers refer to pages)

PRINTED IN U.S.A.